LMS/BR CLASS 7 4-6-0 REBUILDS

The Rebuilt Jubilee, Patriot and Royal Scot Locomotives

LMS/BR CLASS 7 4-6-0 REBUILDS

The Rebuilt Jubilee, Patriot and

Royal Scot Locomotives

David Clarke

THE CROWOOD PRESS

First published in 2014 by
The Crowood Press Ltd
Ramsbury, Marlborough
Wiltshire SN8 2HR

www.crowood.com

British Library Cataloguing-in-Publication Data
A catalogue record for this book is available from the British
Library.

ISBN 978 1 84797 651 2

Photographic Acknowledgements
Photographs are from the author's collection unless stated
otherwise.

Frontispiece: 46129 The Scottish Horse (rebuilt December
1944) is seen at Bletchley on the West Coast Main Line.
A Crewe North engine from November 1957, it then moved
to Carlisle Kingmoor in July 1961, before moving across the
city to Upperby in September 1961. A move back to Crewe
North followed in February 1962, before a final move to
Longsight in September 1962 and withdrawal in June 1964.
Following storage at Longsight, the engine was towed to
Central Wagon Co. at Ince, Wigan, for scrapping in November
1964. The engine has AWS (August 1959) and a speedometer
(November 1960). When it was a Longsight engine, it was
used regularly on the Manchester–Buxton passenger services
in the winter of 1963–4, one of the last regular passenger
workings for the class. (Colour-Rail BRM)

Designed and typeset by Guy Croton Publishing Services,
Tonbridge, Kent

Printed and bound in India by Replika Press Pvt. Ltd.

CONTENTS

Acknowledgements 6

Introduction 7

1 Origins 12

2 Detail Differences 30

3 Liveries, Names and Nameplates 50

4 Testing and Experimentation 74

5 Allocations 82

6 Rebuilt Class 7s to the Rescue – The Severe Winter of 1962–3 130

7 Decline and Withdrawal 132

8 Maintenance of the Locomotives 146

9 Withdrawal and Disposal 152

10 Summary 160

11 Preservation 166

Appendices 174

Recommended Reading 205

Index 206

ACKNOWLEDGEMENTS

Producing a book is not a solitary process; a number of people have provided much help and assistance. Thanks are due to the following photographers for digging through their archives of photographs: Kenneth Tyler (and for his reminiscences of the rebuilds on the Buxton–Manchester services); Peter Groom; Norman Preedy; Malcolm Castledine; and David Cousins. Also thanks to RailOnline (www.rail-online.co.uk) for the use of a number of photographs. To Angela Tarnowski and Mal Siddons of the Sherwood Foresters Museum, who provided access to the various plates used on 46112 *Sherwood Forester*. To Len Pinder, who sorted through a mass of slides and digitally copied them for me and to Steve Taylor for doing some excellent Photoshop work on the colour slides.

I also interviewed as many railwaymen who had direct experience of the class as possible, including: Chris Ward, a fireman at Annesley shed; Alan Newbury, also a fireman at Annesley; Granville Dobson, a fireman at Low Moor shed; and Mike Bentley, a fireman and driver at Buxton and who worked on the engines many, many times.

The published reminiscences of Eric A. Langridge, who was working in the Derby Drawing Office during the 1930s and 1940s, were an invaluable source of inside information. Eric was directly involved with the design of the Stanier locomotives and had a unique view of events and the background as to why things happened. He lived to be well over 100 and published his memories in a number of magazines and books.

Dave Cousins, for digging out some of his black-and-white photographs. Pete Skellon of the Bahamas Locomotive Society provided details of 46115 *Scots Guardsman* during its time when the Society undertook the restoration of the engine in the 1980s and also some of the modifications made to the rebuilt Scots when in BR service.

I also referred to various issues of *Backtrack*, *Steam Days*, *Steam World* and *British Railway Illustrated* magazines, which had reminiscences from former railway staff, again giving valuable insight into the working of these locomotives.

Finally, to my wife Glenis, who has put up with my endless hours of research and for helping with the proofreading.

INTRODUCTION

I am old enough to have seen many of the LMS (London, Midland and Scottish Railway) Class 7 rebuilds in traffic and was fortunate to have been pulled by a number of the locomotives on passenger trains as they were used on 'fill-in' turns from my local station, Trench Crossing between Stafford, Wellington (Shropshire) and Shrewsbury (where they would then work from Shrewsbury to Crewe). As a teenager in 1962, I went on a rock climbing holiday to Snowdonia, necessitating a rail journey from Wellington to Chester and then to Llandudno Junction and down the branch line to Blaenau Ffestiniog, where I left the train at Betws-y-Coed. The train from Chester to Llandudno Junction was hauled by 46163 *Civil Service Rifleman* and the return by 46152 *The King's Dragoon*

Guardsman, both doing the job they were designed for, that of express passenger trains. On one of my visits to Shrewsbury around 1962–3 I was allowed to climb up on the cab of 46125 *3rd Carabinier*, which had arrived from Crewe on a train destined for the route to Bristol and beyond. It was a real privilege to be on the footplate; I only wish I could have stayed on the locomotive on its journey south.

The saddest sighting for me was seeing one of Crewe North's rebuilt Scots, 46155 *The Lancer*, arriving on freight at the military depot at Donnington (on the line between Wellington and Stafford) and, after some shunting, heading off towards Stafford; those 6ft 9in driving wheels were not really designed for mixed freight workings. My last sighting of a rebuilt Class 7

46124 London Scottish *(rebuilt December 1943) is seen at Shrewsbury circa 1959–61 with what is likely to have been a stopping train to Stafford, which was a regular 'fill-in' for an 8A engine that would arrive at Shrewsbury with the 11.45 Manchester (London Rd)–Plymouth, where the engine would come off to be replaced by a Western Region engine. The rebuilt Class 7 would do a return trip to Stafford with three coaches and then work back from Shrewsbury to Crewe with another express passenger train. The overall station roof was removed during 1961–2.*

46148 The Manchester Regiment (rebuilt July 1954) is seen in ex-Works condition at Stafford shed in the early 1960s, when the loco had acquired AWS (May 1959), speedometer drive (October 1960) and the overhead warning flashes (around 1960). Stanier bogie wheels have replaced the Fowler originals. It was common to see rebuilt Class 7s on Stafford shed as they would be used on 'fill-in' turns on the Shrewsbury–Stafford and return stopping services. The author was pulled by 46148 on such a train in 1963. The shed code is difficult to read, but the loco was allocated to Crewe North in April 1962, before moving to Llandudno Junction in September 1962, then on to Holyhead in December 1963. It returned to Llandudno in January 1964, before a final return to Holyhead in April 1964 and withdrawal in November 1964. After storage at Birkenhead shed for a month, it made the long journey to Birds Scrapyard in Morriston, near Swansea, for scrapping in January 1965.

46129 The Scottish Horse (rebuilt December 1944) is seen at Shrewsbury shed in the 1960s. The engine was a long-term resident of Crewe North shed until June 1961, when it was allocated to Carlisle, before being allocated to Crewe again, then going to Longsight Manchester in September 1962. It was withdrawn in June 1964 and scrapped at Central Wagon Company, Wigan, in November 1964. The locomotive has the overhead warning flashes, the additional handrail on the smoke deflectors and the automatic warning system (AWS), but not a speedometer drive. It is fitted with a rocker ash pan (the arm can be seen between the middle and trailing driving wheels), but subsequently in its last years of service it did not have this fitment. G. SHARPE

45526 Morecambe and Heysham (rebuilt February 1947) is seen at Greyrigg on a fitted freight in the autumn of 1964. The engine was allocated to Carlisle Upperby and freight duties were common for its rebuilt Class 7s at this time. The engine has acquired the yellow warning stripe applied at the end of August 1964 and still retains its nameplates and crests above the name. The engine is fitted with AWS (November 1959) and was also fitted with a speedometer (February 1961). The engine had been at Carlisle Upperby since June 1950 and was withdrawn from that shed in October 1964, before being scrapped in February 1965 at one of the scrapyards in the Glasgow area.

A. E. DURRANT AND M. BOAKES

was on 7 February 1965 when visiting Crewe North depot (as part of a Warwickshire rail tour to Crewe Works and depots), where I found one of the last survivors, 46115 *Scots Guardsman*, standing out of steam in fairly shabby external condition and I feared the worst. I subsequently found out that the locomotive was used on a rail tour a week later and that Crewe North cleaned up the engine and made up some nice replica nameplates.

In my eyes, the rebuilt Class 7s had a more powerful look when compared to the far more common LMS Jubilees. They had quite a soft exhaust, but when pushed could produce a real 'bark' from the exhaust that more than matched their contemporaries, the Castles. The class was the mainstay of the West Coast Main Line from Euston to Glasgow; in reality, the London Midland Region needed more Pacifics, but in the absence of these the rebuilt Class 7s would fill the breach and they were regularly overloaded with fourteen-coach trains. They never seemed to hit the high-speed heights of some of their contemporaries, but the majority of the routes they ran on were not conducive to high-speed running. The valve events on the locomotives were specifically designed to operate at a lower speed range, which gave the locomotives 'punch' at lower speeds. This is best illustrated when members of the rebuilt Scots were transferred to Low Moor Bradford and were diagrammed to work a daily

heavy Bradford to Southport semi-fast, usually loaded to ten coaches and with forty stops on a steeply graded route (as far as Manchester). The Low Moor crews were surprised how easy the locos, with their 6ft 9in drivers, could accelerate away with a heavy load and the ability of the boiler to generate massive amounts of steam on the hilly sections.

The LMS rebuilt 7s were nominally in three different classes (the rebuilt Jubilees, the rebuilt Patriots and the rebuilt Royal Scots), but they essentially formed one class, all sharing the same wonderful 2A boiler, but with detail differences both among the classes and within each class. Many of the detail changes between the various sub-types are very subtle and not always easy to spot in photographs. The Engine History Cards (EHC) have been analysed, but as with many other classes, the cards suffer from not being fully updated from the early 1960s onwards (and some engines that were on the North Eastern Region stopped having the changes recorded before others), so many of the later changes (such as speedometer, tender changes and Works visits) were not officially recorded.

There are also some anomalies on the EHCs; as an example, 46101 *Royal Scots Grey* is recorded as having a Light Classified repair between 18 December 1961 and 23 March 1961 and this is the last entry for repairs on the EHC. However, on the same EHC, boiler num-

45535 Sir Herbert Walker, K.C.B. (rebuilt September 1948) is seen on a freight at Carlisle on 18 May 1963, when the engine was allocated to Carlisle Kingmoor and not long before it was withdrawn (in October 1963). The engine had been a long-term resident at Edge Hill, with eight years of service from that depot, and had received an overhaul at Crewe Works in April 1962, before its final move to Carlisle Kingmoor in November 1962. The engine had a long journey for its final disposal, as, following storage at Kingmoor, it travelled to William Rigley's Wagon Works at Bulwell, near Nottingham, where it was scrapped in September 1964. RAILONLINE

ber 12665 was noted as being fitted on 23 March 1962, indicating a heavy repair at this later date. There are at least four other locos where this sort of discrepancy can be identified. In order to identify many of these later changes, I have used an extensive collection of photographs and observations from such as *The Railway Observer*, published by the Railway Travel and Correspondence Society (RCTS).

Where I have used previously published lists for changes and variants I have validated this with photographic evidence and where the photograph contradicts the published data I have not used the published sources. In the course of doing the research for this book I came across a number of contradictory state-

ments, so if there are any errors they are mine not a simple reiteration of 'facts' from an unknown source. The same applies to allocations. The Engine History Cards do not list all of the final allocations, so, for example, the last entry for 46164 *The Artists' Rifleman* is an allocation to Crewe North in 1959; the EHC does not mention the subsequent allocation to 41C Millhouses Sheffield in February 1960 and there are many other examples.

The three separate classes have been well covered in book form compared to many other classes, but the three varieties of Class 7 rebuild locomotives have never been documented and treated as one class, whereas of course from the Operating Departments'

45512 Bunsen (rebuilt July 1948). A magnificent shot taken in June 1964 at Willesden when Bunsen was visiting from its home depot of Carlisle Upperby, to which it had been allocated in May 1949 and where it remained until a move across the city to Carlisle Kingmoor in November 1964. The front bogie wheels are Stanier and the rear set is Fowler, whilst the leading and middle driving wheels are Stanier and the rear set is Fowler, with the usual Stanier-style balance weights. AWS was fitted in November 1959 and speedometer drive in February 1961. The engine survived until March 1965, before scrapping at Motherwell Machinery and Scrap, Wishaw, in July 1965. COLOUR-RAIL

point of view they were the same. What does this book have to add to the story beyond the obvious one of having colour photographs? I have tried to summarize all the detailed changes in a compact format so that when viewing a photograph the reader should be able to identify any changes made to that particular locomotive. The key element of the book is to give a full picture of the locomotives in operation, covering all the depots to which they were allocated and the principal trains that the class worked, as well as the experiences of the driving and fitting staff.

The decline of the class and the allocation to some very unglamorous depots has not been covered in depth before, as well as their widespread use filling in for failed diesels during the early 1960s. The allocation to depots such as Wigan Springs Branch, Saltley, Low Moor and Annesley would appear not to make any sense, until one looks at and understands the traffic workings from those depots.

The photographs have been chosen to show as many of the possible variations through the classes. What surprised me, as I have been collecting photographs and slides of these engines for over thirty years, has been the paucity of good colour photographs, but the recent RailOnline collection has made the task a little easier. I have also tried, where possible, not to use photographs that have been used many times before; not an easy task, but I hope I have succeeded.

CHAPTER ONE

ORIGINS

William Stanier took over the Locomotive Department of the London Midland and Scottish Railway in 1932, bringing with him many of the progressive engineering and design influences from the Great Western Railway at Swindon. The most powerful engines in the LMS fleet at that time were the seventy Royal Scots introduced in 1927. Whilst these proved to be successful machines they were not without fault, principally with axle box failures, rough riding and smokeboxes that proved difficult to keep airtight (this was due to the design of the built-up smokeboxes). Stanier quickly initiated some modifications to the springing and axle boxes as well as the bogies, resulting in a considerable improvement in the reduction of hot boxes, the locomotives having been provided originally with Midland-style axle boxes.

Subsequently, new locomotives were designed and built under Stanier's regime, with the Princess Royal 4-6-2s introduced in 1933 (thirteen locomotives) and the Coronation 4-6-2s introduced in 1937 (thirty-seven locomotives), both classes replacing the original Royal Scots on the heavy principal West Coast trains. In addition, new 4-6-0s such as the Jubilee, introduced in 1934 (187 locomotives), were supplementing the Royal Scots. Prior to Stanier's arrival, Patriots (more popularly known as 'Baby Scots') had been introduced in 1930, the first two incorporating many parts from London and North West Railway (LNWR) Claughton 4-6-0s. Subsequent batches were built from 1932 and Stanier approved the building of more members of the class during 1933 and 1934, the latter batches incorporating some detail changes initiated by Stanier, finishing with a total of fifty locomotives.

So by the end of the 1930s, the LMS had upgraded its passenger fleet, with many engines designed the 'Stanier way'. However, except for the minor modifications initiated by Stanier, the Royal Scots were very much as designed and built in 1927. As a result, they were gradually replaced on the very top services on the West Coast Main Line by the Princess and Coronation Pacifics. The issue of smokeboxes drawing air became more of a problem; the experience of the 'drumhead'-type smokeboxes on the Stanier Jubilees and Black Fives pointed the way forwards. Cracked frames were also increasing on the Royal Scots. The following table (from a report by the LMS engineer E.S. Cox, who worked for Stanier) shows how the incidence of frame fractures had increased to a point where remedial action was urgently required.

The Royal Scots also had an additional operational constraint, in that they were too heavy to be used on

Original Royal Scots: Frame Fractures	
Years of Age	**Average Frame Cracks per Engine Since Construction**
6 (1933)	1
7 (1934)	2
8 (1934)	5
9 (1936)	10
10 (1937)	17
11 (1938)	21
15 (1942)	37

the Midland Main Line. As early as 1934, the Locomotive and Electrical Committee, when discussing the reboilering of the experimental locomotive 'Fury' with a new design of taper boiler, noted that: 'the alterations to include provision of a new boiler of the taper barrel type, which it was proposed to fit to the Royal Scot class as a whole'.

In 1942, Stanier addressed these problems by reporting that twenty of the Royal Scots should be rebuilt with new taper boilers (as per the two rebuilt Jubilees), new cylinders and a new design of smokeboxes. The resultant reduction in weight would also have the bonus of allowing the locos to operate over the Midland Main Line. However, before we move on to the rebuilding process of the Jubilees, Patriots and Royal Scots, we need to consider the development process that concluded with the design of the 2A boiler and the role played by the rebuilding of 6399 'Fury' with its type 2 boiler.

The Rebuilding of the 'Fury' into British Legion

6399 'Fury' had been built in 1930 and was effectively an additional Royal Scot fitted with an experimental ultra high-pressure boiler designed by Schmidt-Henschel and built by the Superheater Company. Steam at 900psi was fed to the middle cylinder, with the exhaust from the middle cylinder being mixed with steam at 250psi for the two outside cylinders. The testing of this unconventional locomotive had only just commenced when on 10 February 1930 one of the high-pressure tubes burst, resulting in the death of the representative of the Superheater Company (Mr Lewis Scofield) and causing injuries to the driver and fireman, as well as the inspector who was also on the footplate.

Following this accident and the results of an investigation, the high-pressure circuit was modified and the

6399 'Fury'. The high-pressure experimental locomotive is seen at the Derby Works before the decision was made to rebuild the locomotive with a more conventional boiler. The locomotive was in essence a Royal Scot, but was fitted with a high-pressure boiler. During testing in February 1930, it suffered a catastrophic blowback on to the footplate. The engine was steamed a number of times for further tests after the accident, but no advantages to this high-pressure boiler were revealed before Stanier made the decision to rebuild the locomotive.

46170 British Legion (rebuilt October 1935) arrives at Rhyl with a short express on 11 July 1958, when the engine was allocated to 1B Camden. The engine became familiar on the North Wales route, with periods spent at Crewe North and Holyhead, and was withdrawn from Llandudno Junction. BLENCOWE COLLECTION

6170 British Legion (rebuilt October 1935) is seen with the nameplate covered up the day before a ceremonial unveiling at Euston station on 12 November 1935, when Earl Jellico unveiled the name. The engine is fitted with the original single chimney and the backwards-sloping outside steam pipes, both of which subsequently changed. The boiler barrel of 6170 is 15in (381mm) longer than the subsequent 2A boilers, hence the sloping outside steam pipes. The front frames above the platform are a different shape to 'Fury', which points to new frames when rebuilt. The steaming of the No.2 boiler was problematic at first, but revisions improved matters considerably and these were incorporated into the 2A boilers fitted to the rest of the rebuilt Class 7 locomotives. The front coupling rods are of the split-brass type.

6170 British Legion. A front view of the engine following rebuilding, but still with the single chimney. The front 'faces' of the different classes that comprised the rebuilt Class 7s were all different. 6170 was also unique in that it was the only rebuilt Class 7 to carry the LMS red livery.

6170 British Legion is seen after its naming ceremony and before the fitting of the double chimney. The wheels and motion are all Fowler and retain the split brasses on the leading crankpins, but in line with the rebuilt Scots these were replaced by the later circular type. The outside backwards-sloping steam pipes have also been replaced; the replacements shown here are vertical and much wider. The engine is also fitted with the BTH-type speedometer drive. These were removed during World War II and later replaced by the standard Smith-Stone type. The vacuum pump previously fitted has now been removed and the top feed was unique to this engine, being fitted close to the dome.

46170 British Legion (rebuilt October 1935) is seen outside the paint shop at Crewe Works in July 1949 and displays its double chimney, which replaced its original single chimney. Revisions to the boiler and the fitting of a double chimney improved the steaming of the engine so that it was similar to the rest of the rebuilt Class 7s. The livery is the LMS 1946, with the addition of British Railways on the tender and the changed cab side number. The BR smokebox looks like the short-lived 'serif' style, which was later changed. The engine still has the split brasses on the leading axle coupling rods. It entered Works for a Heavy Intermediate overhaul on 9 June 1949 and left on 20 July 1949, so at the date of the photograph it had just entered Works. When boiler repairs were required, the engine would spend longer in Works than usual for the other rebuilt Class 7s, as the boiler was not interchangeable and the repaired frames would have to wait for their own boiler to be repaired rather than receiving an already overhauled one. Crewe built nine 'spare' boilers for the rebuilds, which served to keep the repair time down significantly. COLTAS TRUST

locomotive was tested in July 1931 when stationary. Subsequently, a number of stationary tests were conducted before some further road tests were carried out in 1934. The road tests showed that the steaming was poor and there was not the expected improvement in coal consumption. Stanier's report of 1934 to the LMS Board said: 'in spite of elaborate and prolonged trials, [it] had not proved successful, the steaming in every case having been unsatisfactory'. E.A. Langridge, in his recollections, stated that the original intention had been to reboiler 'Fury' with a standard Royal Scot parallel boiler, but with Stanier now at the helm more progressive ideas would prevail and as a result 'Fury' was provided with a new taper boiler, cylinders and single chimney.

The LMS chief draughtsman at the time was the excellent T.F. (Tom) Coleman, who came from the Horwich Works. He would be the man responsible for turning Stanier's outline schemes into practical and effective locomotives. Coleman was an intuitive engineer and because of his Horwich background was not stuck in the Derby or Crewe way of doing things. It was Tom Coleman who had been responsible for overseeing the detailed design of all the Stanier locomotives and although there had been some difficulties initially regarding the amount of superheating in locomotives such as the Black Fives and Jubilees, these subsequently became excellent machines.

Coleman produced a design for the No. 2 boiler for 6399 'Fury' with a sloping throat plate, a regulator in the dome and a barrel with a continuous taper. This boiler resembled a Swindon product more than anything previously designed by the LMS and it showed how the Crewe Drawing Office was keen to adopt Swindon ideas. Interestingly, the detailed boiler was designed by G.R. Nicholson, who came from Horwich, as also did Wilcocks, who designed the cylinders. For this new boiler the grate area was set at 31sq ft (2.9sq m) after various schemes had been considered and the firebox tube plate was recessed into the barrel by 1in (2.54mm), forming a small combustion chamber. The boiler was of a greater diameter than those fitted to the Jubilee; the 1934 LMS Board minutes indicated that the same boiler would be fitted to the rest of the Royal Scots.

Not only did 6399 'Fury' receive a new boiler, but also new cylinders that reflected ideas from the French engineer Chapelon regarding the flow of steam through the cylinders. These cylinders were considered to be a distinct improvement over the original cylinders fitted to the Royal Scots. They were based on the cylinders used on the Jubilee, but with an easing of the sharp bends and combining in one casting the inside cylinder and the smokebox saddle containing all the exhaust passages for the inside and outside cylinders. The front frames on the rebuilt British Legion were a different shape to those fitted to 'Fury', so at least the front frames were replaced, or it is possible that a new set of frames was put under the rebuilt locomotive.

Initial response by crews to 6170 British Legion was positive, with the engine being considered a good puller and fast, if you could hang on. The engine was used regularly before the war on a heavy (fifteen-coach) Birmingham –Euston train and when allocated to Longsight in 1935 was used on the Manchester to Euston expresses.

The 1934 LMS Board minutes said that the Royal Scots would be rebuilt, but the first rebuilds were actually two Jubilees.

The Rebuilt Jubilees

There had been much discussion in the LMS during the late 1930s regarding the provision of motive power and how to improve it. A number of proposals (DE16) were drawn up in 1937; one of them was effectively the No. 2 boiler mounted on a 2-cylinder chassis similar to a Jubilee and fitted with 20in (508mm) cylinders. The LMS had experienced some problems, particularly with inside big ends on the Jubilees, and using 2 cylinders would eliminate this problem. The use of 2 cylinders would also keep the weight down and the proposal was to reboiler the Jubilees and make them a 2-cylinder engine. However, with changes to the Jubilee boiler and upgrades by the Civil Engineering Department the proposal was shelved. Also it would have been expensive to replace the Jubilee boilers, many of them being quite new; it was cheaper to modify the tube layouts and superheating.

The long discussions regarding the Jubilee boilers and their general performance also prompted some research into the internal streamlining of the steam

5735 Phoenix *(built November 1936). This view (taken in 1939) shows the first of the two Jubilees before its rebuilding with the 2A boiler in April 1942. The Works order for the conversion was allocated in October 1939 and specified 2 Jubilee in the 5721-42 series. This would then free up two relatively young Long Firebox boilers back into the pool for the rest of the Jubilees. Even though the engine was relatively new, the cylinders were replaced as well as the new boiler, although the original boiler and cylinders could go into the 'spare' pool.* M. BENTLEY

ports and passages. Tom Coleman persuaded Stanier to go ahead with a scheme of combining the Jubilee chassis and a redesigned No. 2 boiler. The Drawing Office at Derby had entries for the drawings for the conversion of two Jubilees in August and September 1939. During April and May 1939, 6170 *British Legion* was tested against an original Royal Scot and it is supposed that the results of these tests may have had some bearing on the design of the two rebuilt Jubilees. What is clear is that the drawings for the rebuilding of the Royal Scots were already in hand when these tests took place.

Under the supervision of Tom Coleman, the Derby Drawing Office was given the task of producing a revised No. 2 boiler and fitting it to two Jubilees and also fitting the new style of cylinders. The starting point was the boiler designed at Crewe for *British Legion*, but to reduce the length of the boiler barrel by 1ft 3in (381mm). The revised boiler was to be referred to as 2A and was to be one of the finest steam raisers ever produced for a British railway. The new cylinders were based on the new ones fitted to British Legion, but because of differences in frames and so on

the pattern used for *British Legion* had to be modified for the two Jubilee, meaning that if *British Legion* ever required a new middle cylinder one could not be cast!

Coleman then set out a scheme giving the 2A boiler to the Jubilee class, on the basis that some of the early engines in this class would soon require new boilers and that rebuilding two engines out of the later Long Firebox Jubilees would free up two 'spare' boilers that were comparatively young. For the Jubilees, the rebuilding process required the changes detailed below.

A new straight reversing reach rod (due to an increase in boiler diameter over the original) was fitted outside the intermediate splasher and also required a modified steady bracket.

New outside cylinders that were to be fitted were similar to the ones fitted to 6170 *British Legion*, but with small modifications due to the differences in motion centres. The cylinder diameter stayed the same as the original Jubilee at 17in (432mm), as all the motion parts that were retained had been stressed and loaded for that diameter and if the larger 18in (457mm) cylinders had been fitted, more of the motion parts and

A weight diagram for the two rebuilt Jubilees. Every class of locomotive was given a weight diagram, which gave the basic dimensions and axle weights for the class.

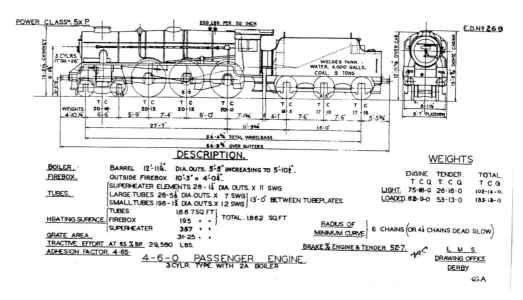

45736 Phoenix (rebuilt April 1942) is seen in its final condition, except for the addition of the yellow warning stripe. The engine has a 5A Crewe North shed plate; the engine was allocated there between September 1960 and September 1962, when it moved to Holyhead. The engine has a welded tender, whilst its sister, Comet, had a riveted tender. When first rebuilt, these two Jubilees were referred to by the Holbeck crews as 'the heavenly twins' – high praise indeed. The mountains of ash in the foreground indicate the running down of steam facilities and the difficulty of attracting shed staff in the early 1960s.

crankpins would have needed replacing to take the extra loads and stresses. The increase in the boiler diameter also meant that the new inside cylinder would have a different radius for the smokebox saddle.

The drawings were issued to Crewe Works in early 1940 for parts to be made and for the boiler shop to build the boilers, which were completed in October 1941. The Works was engaged in war work as well as maintaining the existing fleet, which explains the length of time to produce the two new boilers.

5735 and 5736 were selected based on the mileages since they were last shopped. They had accumulated 130,000 miles (209,200km), which far exceeded others of the specified series that came to Crewe Works for repair. Jubilees were also repaired at Derby and St Rollox, so Crewe did not have a free choice of engines to rebuild.

5736 *Phoenix* was rebuilt with its original driving wheels, which had hollow axles, although all wheel sets were considered interchangeable, so solid axles could be seen on locomotives that previously had hollow ones. Rebuilding also meant new brake hangers and articulated twin brake blocks similar to those fitted to the later builds of the Black Fives (which had this feature from 1938 for new builds).

The original frames were retained, as were the motion and wheels, but some changes were made to the frame, including an extension welded to the top between the centre and trailing driving wheels. A new cab front was also required and the cab front windows were revised, being reduced in size due to the larger boiler. Because the boiler was pitched 4.5in (114mm) higher than the original Jubilee boiler the cab floor was raised by 7in (178mm) to accommodate the increased height of the fire-hole door. This in turn affected the tender shovel plate, which was 5' above the floor, which meant that the tenders for the rebuilt Class 7s had to come from a pool of similar tenders; a standard Jubilee tender would not be suitable.

A double chimney based on the one experimentally fitted to Jubilees 5553 *Canada* and 5742 *Connaught* in 1940 was fitted, but had to be modified to reflect the fact that the boiler was pitched 4.5in (114mm) higher and the smokebox was 5in (127mm) larger in diameter.

5736 *Phoenix* was fitted with a BTH speedometer drive on the left trailing driving axle, but this was probably taken off in 1944 when the LMS removed all its BTH speedometer drives due to spares problems in wartime.

Both rebuilt Jubilees retained the Stanier 4,000gal (18,184ltr) tenders fitted when they entered Crewe Works for rebuilding, both being riveted versions. 5736 *Phoenix* subsequently acquired a welded version of the tender in 1952.

5736 *Phoenix* was released into traffic on 11 April 1942, followed by 5735 Comet on 14 May 1942.

As with 6170 *British Legion*, the two rebuilt Jubilees initially suffered from rough riding. An inspector riding on one of the two rebuilt Jubilees when joining the train asked the driver how he liked the engine. The driver's response was 'rough' and when the inspector pointed out it was only a short time out of Works, the driver maintained 'can't help that, she is still rough'.

The Rebuilt Scots

After the rebuilding of the two Jubilees in 1942, a start was made in 1943 on rebuilding the Royal Scots. This should not be attributed to Stanier, as the work was initiated by C.E. Fairburn when Stanier was seconded to the Ministry of Munitions, although the latter had gained authorization for the work.

The rebuilding process followed the example set by the rebuilding of the two Jubilees. The cylinders were based on the new ones used for the rebuilt Jubilees, but because the Royal Scot motion was retained and all the bearing surfaces were bigger, the cylinders were bored out to 18in (457mm). The same 2A boiler as used for the rebuilt Jubilees was fitted, as well as a new middle cylinder incorporating a new smokebox saddle and new smokebox.

Many of the rebuilt Scots kept the original coupling rods, complete with split brasses in square ends for the leading crankpin, in addition to the coupling rods, crossheads and valve gear. New frames, which were deeper and more rounded at the front buffer beam, were provided, as well as new-style spring hangers and springs. The cab front was modified, although the rest of the original cab was retained.

The original valve spindle crosshead guides, which on the original locomotives were carried on an outrigger from the motion plate, were transferred to the rear

6100 Royal Scot *is seen inside Crewe Works in 1934 after the engine returned from a trip to America. The bell is still in place, but the holes in the smokebox for the headlamp required in America are being filled in. Except for the bell, the engine shows what an original Scot looked like before the extensive rebuilding that replaced the frames, cylinders and boiler. The wheels are of note, as they are fitted with a style of balance weights and webbing on the spokes adjacent to the crankpins usually seen on the later batches of Patriots and the Jubilees introduced in mid-1934.*

46128 The Lovat Scouts *(rebuilt June 1946) is seen at Patricroft shed, Manchester, on 1 December 1960. Rebuilt Class 7s could be seen regularly at Patricroft, as engines coming off trains at Manchester Exchange (from Bangor, Llandudno and so on) would be serviced at the depot, before working back out from Manchester. In later years, Carlisle's rebuilt Class 7s would be seen on fitted freights from Carlisle–Manchester. At the time of the photograph, the engine was allocated to Crewe North (arriving on May 1957), before departing for Carlisle Upperby in August 1962. The AWS was fitted in February 1959 and a speedometer in May 1960. The driving wheels are Fowler originals, with the bogie wheels being Stanier.* COLTAS TRUST

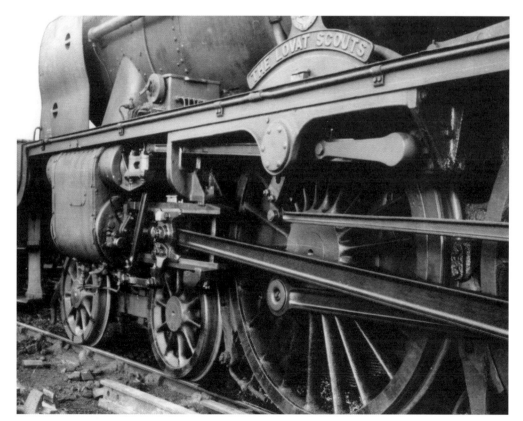

46100 Royal Scot (rebuilt June 1950) is seen under repair in 2012, displaying the new frames fitted when it was rebuilt. The frames and chassis parts were to the latest thinking of the LMS and show the reinforcing plates fitted over the frames around the axle horn guide. The locomotive is currently having major rectification work done at Crewe, following its overhaul at Southall. DAVID BURTON

46100 Royal Scot. This shows in detail how the smoke deflectors, which were fitted from 1950 to 1952 to all the rebuilt Class 7s, were attached to the engine, plus the handhold cups to assist staff climbing on the locomotive. DAVID BURTON

46100 Royal Scot. This shows the front of the engine whilst under repair at Crewe. The front of a rebuilt Scot is substantially different to that of a rebuilt Patriot. DAVID BURTON

46100 Royal Scot *has been
restored in LMS red livery, which
none of the rebuilt class actually
carried in service, but it makes an
interesting change to BR green. The
engine was only briefly in service
following restoration, but a fire whilst
being transported and a major
problem with the axle boxes have
meant that the loco has been in a
dismantled state for a number of
years. The large backing plate for
the oversize nameplate can be
clearly seen.*

DAVID BURTON

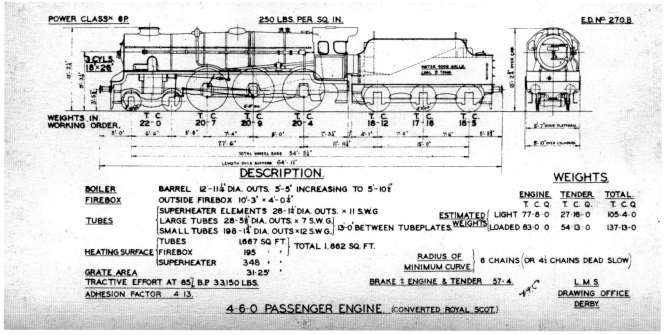

Weight diagram for the rebuilt Scots.

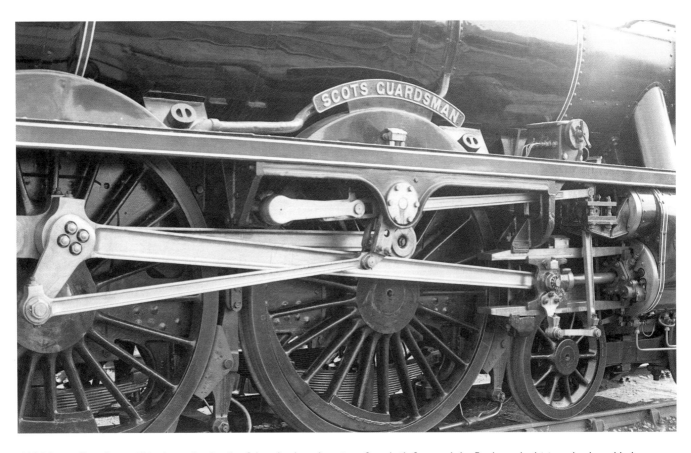

6115 Scots Guardsman. This shows the details of the wheels and motion of a rebuilt Scot and the Fowler-style driving wheels and balance weights, all transferred across from the original Royal Scots. The engine has fluted coupling rods and the frames show the later type of spring hangers fitted during its rebuilding.

6115 Scots Guardsman. This shows the new cylinders fitted when the engines were rebuilt. The cylinders were based on those fitted to the Jubilees, but with internal changes to improve the steam flow and revised valve events from the original Royal Scots. Al the motion was reused from the original Scots, but was modified for the fitting of the piston valve slide bar arrangement. The Stanier bogie wheels clearly show the bevel rim fitted to these wheels and the reduced diameter of the centre boss compared to the Fowler type. The later circular locating pin for the front crankpin can also be seen.

6103 Royal Scots Fusilier *(rebuilt June 1943). The first rebuilt Scot is seen after its conversion, painted in workshop grey, although the LMS lettering has been applied. The locomotive was eventually painted plain black with LMS lettering. The photograph shows the absence of the large sandbox on the platform subsequently fitted to all rebuilt Scots and also shows the rectangular split brasses fitted to the front driving-wheel crankpins; these were gradually replaced with a more modern circular fitting. The engine was unique in being fitted with the BTH speedometer, which was removed shortly afterwards in 1944 due to spare parts becoming in short supply and has yet to receive its smoke deflectors (not fitted until around 1950–51). The top feed cover has a flat top, which was subsequently modified with a curved top, and the bogie wheels were of the Fowler type fitted to the original Royal Scots.*

The engine, following rebuilding, was allocated to Leeds Holbeck (from August 1943) and remained at Holbeck until October 1958, when it was allocated to 14B Kentish Town, followed by transfers to Trafford Park, Kentish Town (again), Saltley and Carlisle Upperby. It returned for a brief stay at Leeds Holbeck, arriving in September 1962, before being withdrawn at the end of December 1962. The locomotive was stored at Leeds Farnley Junction between January and August 1963, before being scrapped at Crewe Works on September 1963.

valve chest covers and the outriggers were cut back.

Because the sanders on the Scots were both in front and at the rear of the centre drivers (unlike the rebuilt Jubilees), the original sandboxes under the platform were retained. However, it was quickly discovered that additional sanding capacity was required and after the first few rebuilt Scots had been converted a larger capacity sandbox was placed on the platform with two separate filling lids. The first few Scots already rebuilt had this new-style sandbox retrospectively applied.

The first rebuilt Scot to emerge was 6103 *Royal Scots Fusilier* in June 1943, closely followed by 6109

Royal Engineer in July 1943, and for the next few years Crewe concentrated on rebuilding Royal Scots. However, in 1946 the first of the Patriots was rebuilt and for four years Patriots and Royal Scots were rebuilt in parallel with each other, until the rebuilding process concentrated on finishing the remaining original Scots. However, this was a protracted process, with the last original Scot, 46137 *The Prince of Wales's Volunteers* (South Lancashire), not being rebuilt until 26 March 1955, when some of the early rebuilds were already twelve years old.

46146 The Rifle Brigade (rebuilt October 1943) is seen sporting its new BR livery and smoke deflectors circa 1949–50. The smoke deflectors had been fitted in October 1949 and the engine has its 5A Crewe North shed plate. It left for Carlisle Upperby in July 1951. The engine still has the rectangular bearing on the front coupling rod and the original style of top feed cover. The handhold above the front steps is in the original position on the platform and was later moved to the bottom of the smoke deflector. The bogie wheels are Stanier-style with a bevel rim. AWS was fitted in November 1959 and the speedometer in November 1960.

6126 The Royal Army Service Corps (rebuilt June 1945) is seen after its rebuild, displaying its plain black LMS livery with shaded yellow lettering. The coupling rods are the later type with a circular fitting on the leading axle.

The Rebuilt Patriots (Baby Scots)

Having initiated the rebuilding of the Royal Scots, which would continue in 1945, the LMS authorized eighteen Patriots to be part of the post-war rebuilding process. Job No.5390 issued in August 1945 said: 'To meet post-war needs of accelerated express passenger services it has been decided to convert eighteen 5X parallel engines to take taper boilers; eight boilers in 1946 Programme. Engines to be selected when requiring boiler overhaul. To have 4,000 gallon tenders.' The first twelve Patriots could not be considered for rebuilding as they were considered 'non-standard'. The later Patriots incorporated some Stanier ideas and details, including wheels, spring hangers and so on.

Rebuilding the original Patriots made more sense than rebuilding further Jubilees, as the Jubilees were a more modern machine with better smokeboxes, whereas the Patriots suffered from smokebox leakage problems in the same way as the original Royal Scots. The first rebuilt Patriot was 45521 *Rhyl* in October 1946, closely followed by 45530 *Sir Frank Ree* a week or two later, which also meant that these engines were out-shopped in the new LMS 1946 livery. As the rebuilding of the eighteen Patriots took place over a relatively short period, many of them were out-shopped in the LMS 1946 livery.

The changes in the rebuild process were as outlined below and followed much of what had been done for the two rebuilt Jubilees and the rebuilding of the Royal Scots.

A new 2A sloping throat-plate boiler, with a working pressure of 250psi, was fitted. A rocking grate and hopper ash pan were also fitted; the hopper door-operating mechanism could be seen between the middle and trailing driving wheels on the driver's side. The fitting of rocking grates had started with a batch of Black Fives in 1945 and was part of the drive to

5521 Rhyl *(built March 1933). This shows a Patriot in its original condition. It has the Fowler-style wheels, but with Stanier balance weights that were carried forwards into the rebuilding process. Like the other rebuilds, much of the engine was replaced, as well as frames, cylinders and the boiler. The tenders were also replaced, as all the rebuilt Patriots were given Stanier 4,000gal (18,184ltr) tenders, usually welded, although some rebuilt Patriots did run with Stanier 4,000gal riveted tenders.*

reduce the amount of effort to dispose and prepare engines, thus making life easier for the crews. This arrangement was also used on the post-war rebuilding of the Royal Scots, but it is not known how many of the already rebuilt Scots had the ash pans revised for the rocking grate. Certainly some were not fitted, as Mike Bentley reported that near the end of its life 46129 *Scottish Horse* did not have a rocking grate and hopper ash pan.

A new circular type of smokebox to replace the existing flat bottom type with self-cleaning plates was fitted. A new smokebox saddle correctly radiused for the new 2A boiler would also be required and fastened to the top of the new inside cylinder.

As per the rebuilt Jubilees, new 17in (432mm) cylinders were provided. These had improvements to the steam circuit and streamlining of the ports. Eric Langridge suggested that this diameter was decided upon rather than the 18in (457mm) cylinders for the Royal Scots because the original motion would be retained and as the boiler pressure was higher than

that of the original Patriots (225psi compared to 200psi), the loading and stresses on the original motion parts would be fine with a 17in-diameter cylinder, but not with an 18in.

Again as per the rebuilt Jubilees and rebuilt Scots, the same double chimney and blast pipe were fitted. However, because the existing Fowler cab was too narrow, a new cab in the style of the Jubilees with double windows was fitted.

Even though the later-build original Patriots were not that old, it was decided that new frames incorporating all the latest LMS improvements to design, materials and construction would be provided. The design of spring hangers and springs had been revised on the later-build Black Fives and these changes had been incorporated into the rebuilt Jubilees and Scots. However, some notes say that the first eight Patriot conversions only had new front frames, as some of the original Patriots had already received new frames before the rebuilding process started. The 1948 and 1949 conversions certainly had new frames.

45529 Stephenson (rebuilt July 1947). When rebuilt, the engine was unnamed, but carried the backing plate for a nameplate until it was named in July 1948 to commemorate the 100th anniversary of George Stevenson's death. The engine was in Crewe Works between 8 July and 7 August 1948 and would have acquired the BR mixed traffic livery and 'British Railways' on the tender. The engine was displayed at Chesterfield goods yard for the commemorations of Stevenson's death (12 August 1848), Chesterfield being his last place of residence and where he is buried in Holy Trinity Church. The engine has the early type of top feed cover with the flat top; the wheels are all of the Stanier type. R.J. BUCKLEY

45527 Southport *(rebuilt September 1948) is seen at Polmadie in March 1950. It is a long way from its home depot of Edge Hill. The engine is in the early British Railways livery of black and is lined out in the LNWR style. This livery was later only used for mixed-traffic engines; the passenger engines were repainted BR green. The engine has the later style of top feed and received its smoke deflectors in November 1951.*

Along with the new frames, new brake hangers with articulated twin brake blocks were fitted, again following the design of the later Black Fives. The sand pipes were different from the rebuilt Scots, as the Patriots only had sanders on the front of the middle drivers, whilst the Scots had sanders at the front and back of the middle set of driving wheels, hence the need on the rebuilt Scots for the large sandbox on the platform with the two fillers. The rebuilt Patriots retained the original sandboxes below the platform.

Unlike the rebuilt Scots, which were already fitted with Stanier 4,000gal (18,184ltr) tenders, the Patriots were fitted with the original Fowler 3,500gal (15,911ltr) tenders and would require new 4,000gal (18,184ltr) welded tenders. These were intended to be moved from Jubilees, although in fact most were built from new. As with all things LMS, some of the rebuilt Patriots received riveted versions of the tenders following Works visits. The wheelbase of the bogies was 6ft 3in (1,905mm), not the 6ft 6in (1,981mm) given in the Irwell Press book about the Patriots (see Recommended Reading, below).

The cost of the later conversions was given as £4,800 against the cost of the 1946 conversions at £4,150; the difference in cost could be attributed to the reuse of parts of the frames. The rebuilding of the Patriots was suspended with only eighteen examples being converted – the new BR CME, R.A. Riddles, overruled a request for further rebuilds on the basis that any new

Class 7 power would be covered by the new Britannias. The last Patriot rebuilt was 45522 *Prestatyn*, which left Crewe Works on 7 February 1949.

Once the rebuilding process for the Royal Scots and Patriots had started, Crewe, in its inimitable fashion, organized the process to be as efficient as possible. Peter Rowledge, who was at Crewe Works from 1951 to 1956, outlined the process for the remaining Royal Scots as follows:

- A new set of frames with new cylinders, spring hangers and springs would be erected in the Heavy Machine Shop using a few renovated components.
- A new 2A boiler would be built in the Old Works boiler shop and upon completion would be moved to the Boiler Mounting Shop, where all the ancillary boiler fittings would be mounted and on completion the completed boiler would then be taken to No. 10 shop.
- The new frames would then join the boiler in No. 10 shop, where the frames and boiler would wait (sometimes for several weeks) before the next engine selected for conversion would arrive. The selection of which locomotive to convert was based on whether the boiler needed replacing.
- Once the selected locomotive arrived, the conversion process could commence. The only components reused for the Royal Scots were the driving wheels, bogie wheels, motion, cab and some minor components. The Royal Scots had already acquired Stanier

45530 Sir Frank Ree (rebuilt October 1946) is seen at Willesden, where it was a bit of a 'pet', the depot keeping the engine in excellent external condition. It was one of the last rebuilt Class 7s to go through Crewe Works in August 1963 and was not withdrawn until December 1965 from Carlisle Kingmoor. Careful observation reveals that the nameplate has not been bolted on correctly; at Willesden the nameplates would be removed when the engine was stored and reinstated when it returned to traffic.

45526 Morecambe and Heysham (rebuilt February 1947) runs up the West Coast Main Line in 1961 near Winwick with a fast freight. The rebuilt Class 7s were regularly seen on fast freights, such as the meat and fish trains to London, before they began to lose their passenger workings. 45526 was a long-term resident of Carlisle Upperby and was withdrawn in October 1964. The photographer, the late Jim Carter, was in the author's opinion one of the finest railway photographers and used his unique access as a fireman and driver at Patricroft to best effect.
RAILONLINE

Summary of the Rebuilding Process

Class	1935	1942	1943	1944	1945	1946	1947	1948	1949	1950	1951	1952	1953	1954	1955	Total
Jubilee		2														2
Patriot						2	6	9	1							18
Scot	1		9	9	11	9	4	5	6	6	2	2	4	2	1	71
Total	**1**	**2**	**9**	**9**	**11**	**11**	**10**	**14**	**7**	**6**	**2**	**2**	**4**	**2**	**1**	**91**

Summary of Dimensional Differences

	Cylinders	Weight	Boiler Pressure
Original Jubilee	17 _ 26	79tons 11cwt	225
Rebuilt Jubilee	17 _ 26	82tons	250
Original Patriot	18 _ 26	80tons 15cwt	200
Rebuilt Patriot	17 _ 26	82tons	250
Original Royal Scot	18 _ 26	84.9tons	250
Rebuilt Scot	18 _ 26	83tons	250
British Legion	18 _ 26	84tons 1cwt	225

4,000gal (18,184ltr) tenders, so new tenders were not required as part of the conversion process (unlike the rebuilding of the Patriots). Any rebuilt Patriots would, of course, require a new tender to replace the Fowler type originally fitted.

The rebuilding process meant that every year the railway would receive a number of 'improved' engines, with the peak year being 1948, when fourteen engines were rebuilt. The table summarizes the rebuilding process.

The other table summarizes the small dimensional differences among the four groups of engines. What is obvious is how close the basic dimensions are between the classes; the real differences were in the details.

46104 Scottish Borderer (rebuilt March 1946) sits on shed sometime in the late 1950s, as the engine is yet to acquire AWS or speedometer. The large sandbox with two separate lids for providing sand to the front of the rear driving wheels and the rear of the middle drivers can be seen. The engine was a long-term resident of Polmadie, being allocated there before rebuilding and remaining there until withdrawal in December 1962. The engine, like its fellow shed mates at Polmadie, spent a long time stored after withdrawal and was not scrapped until May 1964 at McWilliams at Shettlestone, near Glasgow. ERIC BLAKEY, TRANSPORT TREASURY

CHAPTER TWO

DETAIL DIFFERENCES

The ex-LMS rebuilt Class 7s nominally belong to four separate classes (rebuilt Jubilees, Patriots, Royal Scots and British Legions), but from an operating point of view they were the same class of locomotive, having in common the magnificent type 2A boiler and sharing many of the chassis components. So, for example, the rebuilt Jubilees and Patriots had far more in common with other rebuilt 7s than they did with their nominal classmates in their originating classes.

However, because of the different origins and the long period of time between the rebuilding of 'Fury' and the later rebuilt Scots, there is a host of detail differences among not only the rebuilt Patriots, Scots and Jubilees, but also among class members.

Wheels

During the thirteen-year rebuilding process, the design of the wheels changed, culminating in the final Stanier design with a bevelled rim and bolted-on balance weights. However, to get there a number of changes were made.

The original Fowler driving wheels fitted to the Royal Scots had cast-in balance weights, which were retained on the rebuilding of the class, although a small number of locomotives received wheel sets previously fitted to the Patriots or Jubilees. The Royal Scots also had Fowler bogie wheels, which had a larger boss at the centre than the equivalent Stanier versions; the Stanier version also had a bevel on the inside edge of the rim. When the original 6100 *Royal Scot* was prepared for its trip to America it was seen with a set of Fowler-style driving wheels, but with balance weights that looked like those that appeared on the later Stanier-style wheels.

The original Patriots also had different sets of wheels; the first two (45500 and 45501) had reused Claughton driving wheels with a large circular boss at the centre of the wheel. The next batch of Patriots had Fowler-style wheels, but with large separate balance weights bolted on, similar to the later Stanier wheels. The later batch of Patriots built under Stanier (but basically as designed by Fowler, with small modifications such as the spring hangers and springs) had wheels similar to the Jubilees with a webbing on the spokes adjacent to the crankpin. Some engines were also observed with the space between the two spokes projecting from the crank filled with a sheet of metal.

The last set of original Patriots was fitted with Stanier wheels with bevelled rim and the bolted-on balance weights. The two rebuilt Jubilees were fitted with Stanier driving wheels with the large separate balance weights riveted through the wheels.

However, as with many of the details with these locomotives, and given the interchangeability of components, there were inevitably changes to the general rules above.

Many of the rebuilt Scots had the following variations:

- Stanier-style bogie wheels replacing the Fowler-style ones, sometimes with a bogie having one axle of each type
- Stanier-style balance weights being fitted to one or more axles, the most common variant being the leading drivers
- Stanier driving wheels and Stanier balance weights

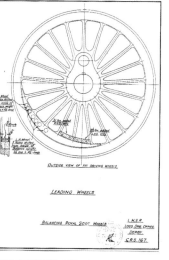

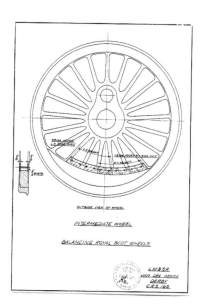

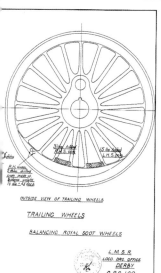

Original Patriot 5516 The Bedfordshire and Hertfordshire Regiment, *built in October 1932, shows the Fowler-style wheels, but fitted with Stanier balance weights bolted through the spokes. This style of driving wheel could be seen on some of the rebuilt Patriots (such as 45545 Planet) and some of the rebuilt Scots. The later-build Patriots had the Stanier driving wheel with a visible bevel to the rim and also balance weights bolted through the spokes. The photograph also shows the Fowler bogie wheels, which had a larger diameter centre boss compared to the Stanier style, which also had a bevel inner rim.*

Royal Scot wheels. The Royal Scots in both original and in rebuilt condition were subject to continual improvements to the balancing of the wheels to improve ride and to reduce any 'hammer blow' to the track. The diagrams here were issued in 1932 and show where additional material could be added over and above the cast-in balance weights. These additional weights were added inside plates (shown shaded), which had rivets from the front to the back to secure them to the wheels. In some cases, it was necessary to drill holes through the wheel to reduce weight at certain points.

Original Royal Scot 46134 The Cheshire Regiment *is seen 29 June 1952 before it was rebuilt (in December 1953) and shows the original cylinders and motion, as well as the Fowler wheels. The engine still has the split brasses to the front driving-wheel coupling rod, a feature that was disappearing on the rebuilds. The original style of fixtures for the driving-wheel springs can also be seen; these were changed when the engine was rebuilt.* E.R. MORTON

46130 The West Yorkshire Regiment (rebuilt December 1949). There had been many attempts to improve the ride and balancing on the Royal Scots, both in original and rebuilt condition, and this view shows additional holes drilled in the balance weights as part of the balancing process.

being fitted to one or more axles; in the case of the leading axle it is assumed that the Stanier wheels were pressed on to the original crank axle
- later-style Fowler wheels with webbing behind the spokes around the crankpin being fitted; this style of wheel was seen on the later builds of the Patriots. 6117 *Welsh Guardsman* was seen with a set of this style of wheel as its leading drivers soon after rebuilding.

In some cases, by examining photographs it is not always possible to see what style of wheel is fitted, but only that the balance weights are of the Stanier style. Examples include:

- 46109 *Royal Engineer* had Stanier-pattern drivers to the leading axle (circa 1960) and built-up balance weights on that axle.
- 46114 *Coldstream Guardsman* had Stanier wheels on all axles in the mid-1950s, but the wheels had been changed back to the 'normal' Fowler style by withdrawal in October 1963.

- 6117 *Welsh Guardsman* was seen in the 1940s with leading driving wheels that appear to be the intermediate-style Patriot wheels, which have the webbing around the spokes adjacent to the crankpin and the Stanier-style bolted-on balance weights.
- 46123 *Royal Irish Fusilier* was seen fitted with Stanier wheels and built-up balance weights to the leading driving axle. The loco has AWS, so must be post-1959. It also has a mixed pair of Fowler and Stanier bogie wheels.
- 46125 *3rd Carabinier* was seen with one pair of Stanier wheels and one pair of Fowler wheels on the bogie.
- 46130 *The West Yorkshire Regiment* had the original Fowler type in 1959, but by 1963 it had Staniers and built-up balance weights on the leading drivers.
- 46131 *The Royal Warwickshire Regiment* was seen with Stanier bevel rims and riveted balance weights on the leading and trailing drivers in 1961. The middle drivers were the standard Fowler ones.

- 46135 *The East Lancaster Regiment* was seen sometime after 1959 with a pair of Stanier bevel rim drivers and the riveted style of balance weights on the leading driving wheels.

- 46141 *The North Staffordshire Regiment* was seen in 1958 with Stanier bevel rim wheels and balance weights on the rear drivers. The centre drivers looked to be Fowler, but with Stanier-style riveted balance weights. The leading drivers were Fowler, but with balance weights in the Stanier style, although not riveted. This unusual combination was still on the engine in 1963 when it was withdrawn.

- 46142 *The York and Lancaster Regiment* was seen with Stanier bevel rimmed wheels and riveted-style balance weights on the leading drivers in the early 1960s. However, by October 1963 the leading drivers had reverted to the original Fowler type and balance weights.

- 6147 *The Northampton Regiment* was seen in the 1940s in LMS livery with the leading drivers having the Stanier-style balance weights.

- 46156 *The South Wales Borderer* (6J Holyhead) was seen in 1962 with the rear drivers being the Stanier type with bevelled rims and the riveted Stanier balance weights; it retained these wheels until withdrawal in 1964. Its leading drivers were of the Fowler type, but fitted with Stanier-style riveted balance weights.

- 46160 *Queen Victoria's Rifleman* was seen in the mid-1950s with Stanier bevel rims and riveted balance weights on the leading drivers.

- 46161 *King's Own* had Stanier wheels with a bevel rim and riveted balance weights on the leading drivers, certainly from 1959.

- 46169 *The Boy Scout* was seen with Stanier-style riveted balance weights on the leading drivers. These remained on the engine until withdrawal in August 1963.

Stanier driving wheels. This shows the final type of Stanier driving wheel, distinguished by the bevelled rim and the use of plates to the front and back of the spokes (with a long bolt to hold them on) and inside, to which lead could be added to balance the wheels. The rebuilt Jubilees, most of the rebuilt Patriots and some of the rebuilt Scots had this type of wheel. As the original Patriots were built with Fowler wheels, there were developments of the Fowler wheel on the Patriots before this final type was fitted to the Jubilees, Black and Fives and so on.

Some of the rebuilt Patriots ran with a full set of Stanier driving wheels, but with Fowler-style bogie wheels; 45521 *Rhyl* was seen in the early 1960s with this combination. Also some of the original Patriots that were built with Stanier driving wheels subsequently acquired Fowler-style wheels on rebuilding, such as 45525 *Colwyn Bay*. 45512 *Bunsen* upon rebuilding had a mixed set of wheels, with the leading and trailing axles having Stanier wheels, but the centre set being Fowler style, with the space between two spokes completely filled in. Subsequently, just before withdrawal in 1965, 45512 *Bunsen* was seen with the Fowler-style Patriot wheels on the leading axle.

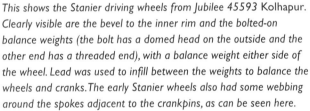

This shows the Stanier driving wheels from Jubilee 45593 Kolhapur. Clearly visible are the bevel to the inner rim and the bolted-on balance weights (the bolt has a domed head on the outside and the other end has a threaded end), with a balance weight either side of the wheel. Lead was used to infill between the weights to balance the wheels and cranks. The early Stanier wheels also had some webbing around the spokes adjacent to the crankpins, as can be seen here.

46156 The South Wales Borderer (rebuilt May 1954). This shows the rear driving wheels on 46156, which have the bevelled rims and balance weights of the Stanier-style wheels. The leading and middle driving wheels remain of the original Fowler type, which most of the rebuilt Scots kept until withdrawal. The engine was at the time of the photograph allocated to Annesley, where it had arrived in November 1963 (from Willesden); it remained there until withdrawal in October 1964. The engine has the AWS air tank, which had been fitted in November 1959. *RAILONLINE*

46109 Royal Engineer (rebuilt July 1943) is seen here sometime before 1960, as the engine does not have the overhead warning flashes. The middle and trailing wheels are of the original Fowler style; however, the leading driving wheels are of the style fitted to the original Patriots, but with Stanier balance weights; they do not have the bevel rims of the later Stanier wheels. The engine is not fitted with AWS, although this was subsequently fitted. The photograph also shows the double sand pipes to the front and rear of the centre driving wheels, which was unique to the rebuilt Scots. The engine was a long-time resident of Leeds Holbeck, before a move to Low Moor in September 1961, then back to Holbeck in June 1962, before withdrawal in December 1962.

46131 The Royal Warwickshire Regiment (rebuilt October 1944) is seen at Holbeck shed being prepared by its crew for its next run. The engine has a 41C Sheffield Millhouses shed plate, which was not listed on the Engine History Card, although other sources indicate it was allocated to 41C in 1960, when a number of rebuilt 7s were allocated to the Midland Main Line, including Millhouses. It stayed until February 1962, when it moved to Llandudno junction. The engine has replacement driving wheels, with both the leading and trailing driving wheels being Stanier with bevel rims and Stanier-style balance weights. A.G. Ellis, collection M. BOAKES

Coupling Rods

The rebuilt Patriots had plain rods as per the original locomotives and these were retained throughout their lives. Similarly, the two rebuilt Jubilees had plain rods when rebuilt and these were also retained throughout their lives. However, when it comes to the rebuilt Scots there were a number of variations as follows.

Rebuilt Scots

A report in March 1945 highlighted that some of the newly rebuilt Scots had been bending coupling rods, which led to changes to the sanding arrangements, but also to changes to the coupling rods. The Mechanical and Electrical Engineering Committee proposed: 'as and when the coupling rods required renewal in the ordinary course of maintenance to provide a stronger design of rod by increasing the cross section and to fit a solid bush at the leading crankpin in place of the

older design of split bush'. This, of course, took longer than one would have anticipated, as engines rebuilt in 1947, such as 6135 *The East Lancashire Regiment*, was out-shopped complete with the early style of split bush to the leading crankpin.

The original Scot rods were fluted, but some locomotives received plain section rods. Locomotives so fitted include:

- 46103 *Royal Scots Fusilier*
- 46109 *Royal Engineer*
- 46119 *Lancashire Fusilier* (seen with plain rods in 1949, but with fluted rods in the mid-1950s)
- 46127 *Old Contemptibles*
- 46131 *The Royal Warwickshire Regiment*
- 46133 *The Green Howards*
- 46138 *The London Irish Rifleman*
- 46160 *Queen Victoria's Rifleman*.

However, as the rods could be moved around at Works visits, locos so fitted with these rectangular rods could

6160 Queen Victoria's Rifleman (rebuilt February 1945) is seen in 1947 after allocation to 9A Longsight and is in its plain LMS black livery. It was renumbered 46160 in September 1948. The locomotive is fitted with plain coupling rods, which were fitted to a small number of the rebuilt Scots before they were fitted with a revised set of fluted rods.

46140 King's Royal Rifle Corps (rebuilt May 1952) is seen at Crewe North shed after a visit to Crewe Works and a repaint. The engine was allocated to Kentish Town between September 1959 and September 1961 (with a transfer to Nottingham for a few weeks in November 1959). The engine received a Heavy Intermediate in June 1960 and a Heavy General in June 1961, becoming one of the last operational rebuilt Scots before its withdrawal at Carlisle Kingmoor in November 1965. It was fitted with AWS in April 1959 and a speedometer drive in June 1960. Interestingly, the crest above the name has gone missing and the space has been painted green; a number of these crests had gone missing on various engines.
G. SHARPE

lose them and gain the original fluted rods again; 46119, 46127, 46133, 4613 and 46160 all reverted to the original style of rods. The last type of coupling rods were fluted, but of a lightweight style due to a different type of material.

The original Scot rods had the front bearing of the coupling rods of a type that split and the coupling rod had a rectangular housing to suit the bearing type. Some of the rebuilt locomotives acquired a more modern pressed-in bearing with rounded ends upon rebuilding, but some were rebuilt with the original split bearings. However, as the years progressed these split-type bearings for the front crankpin were replaced with the circular type.

Cabs

The cabs were different as they reflected the origins of the engines, but also the constraints imposed by the reboilering process. When the Patriots were rebuilt they were given Stanier-style cabs with side windows, as the original Fowler cabs were too narrow for the new 2A boiler. The two rebuilt Jubilees already had Stanier side window cabs and retained these after rebuilding, although a new cab front was required. The Royal Scots had Fowler cabs, but these were big enough to be retained after the rebuilding process.

Crews generally preferred the rebuilt Patriots and rebuilt Jubilee cabs, as they gave better protection in windy, rainy weather, particularly when the wind was hitting the cab side on. For a short period, a number of rebuilt Class 7s had a small bracket put on the outside of the cab side to insert the driver's name to tell passengers who was at the helm. This bracket was located above the number '4' and below the cab side window. This facility had also existed on the LMS in the 1930s, but had since been dropped; the BR scheme also fell into disuse, as drivers were not keen on putting their names on the cab side just in case the train was late and passengers could name names when complaining! Despite analysing hundreds of photos of rebuilt Class 7s, the author has been unable to find a single photo showing a name placed in its holder. These small brackets could be seen on engines through to withdrawal such as 45531 *Sir Frederick Harrison* seen in 1964.

One of the main complaints about the rebuilt Scots was that the cab roof had a transverse joint that dripped water down the backs of footplatemen's necks. The joint was there to facilitate the removal of the rear of the cab roof to provide access to the lifting points at the rear of the engine.

Brake Shoes and Hangers

The Patriots when originally built had single-block brake shoes, but when rebuilt these were changed for a twin-articulated shoe per wheel. These were similar to the twin-articulated brake shoes introduced on the Black Fives from No. 5452 – all the Black Fives from 1938 onwards had the twin-articulated brake shoe. Similarly, the two rebuilt Jubilees also had the upgraded twin-articulated brake, replacing the original single brake shoe of the Jubilees. But just to show that not everything was the same, the rebuilt Scots retained their original single-block brake shoe arrangement.

Front Buffer Beams and Front Platforms

There were differences between the front buffer beams and the depth of the platform drop down to the buffer beams that reflected the different origins of the rebuilt Scots and Patriots. The buffer beams of the rebuilt Scot were deeper and had the riveting countersunk, whilst the Patriots had the rivet heads showing. The difference in buffer-beam heights was caused by variations in the drop down from the main platform to the buffer beam. The buffer-beam shapes were also different, with the result that the cut-out shape was not the same between the two classes.

The cover over the middle cylinder and the shape of the frames above the front platform was also different.

Smoke Deflectors

When the first rebuilds took place the engines were turned out without smoke deflectors, despite the fact

46111 Grenadier Guardsman (rebuilt Jan 1953) and 45527 Southport (rebuilt July 1948) are shown side by side to highlight the differences between the fronts of the rebuilt Patriots and the rebuilt Scots. The frames are different, with the Scots having a curve where the frame meets the buffer beam. The buffer beams are different (one being riveted and one not) and are also of different heights. The AWS 'bash plate' below the buffer beam is to prevent the front coupling damaging the contact shoe attached to the front of the bogie. At the time of the photo, 46111 Grenadier Guardsman had been withdrawn, hence it has no shed plate and the smokebox door handle is missing.

that the original Patriots and Royal Scots were fitted with them. There were many complaints from crews about drifting smoke and the senior drivers at Bushbury made an official complaint, leading to their newly rebuilt locomotives being transferred away. So it was decided to fit deflectors and the first engine to be fitted was 6115 *Scots Guardsman* when it was rebuilt in August 1947. Subsequently, all the members of the rebuilt Jubilees, Patriots and Scots were fitted from 1949, with the work being completed in early 1953. Given the level of complaints regarding drifting smoke, it was a surprise that the process took so long.

Inspector Powell had carried out tests on 6115 *Scots Guardsman* following the fitting of the smoke deflectors and considered them completely ineffective, but his report was ignored by the Operating Department and the rest of the class (as well as the rebuilt Patriots and Jubilees) were subsequently fitted. In Powell's view, the sloping back of the front edge of the deflectors reduced their effectiveness and indeed in July 1954, 46106 *Gordon Highlander* was fitted with a different set of deflectors that were similar in style to those fitted to the original Patriots and also the BR Britannias (which were introduced in 1951) and the 9F 2-10-0s (introduced in 1954). It is not known to the author why 46106 acquired these deflectors, nor why no other members of the rebuilt Class 7s were similarly fitted.

A short handrail had been positioned on the top of the platform just behind the front buffer beam, so that when engine men were climbing up the front steps there was an additional handhold. However, once the smoke deflectors were fitted this handrail was moved to the side of the deflectors, but the process seemed fairly leisurely as some locomotives did not receive deflectors until 1953, so the handrail could not move until after deflectors were fitted.

Rocking Grates

The LMS had been exposed to different approaches for the preparation and disposal of locomotives when 390 of the USA-built 2-8-0s were operating in the UK during World War II, before they were shipped over to Europe after the invasion in 1944. These engines had rocker-type ash grates and self-cleaning smokeboxes, features that were a particular attraction when there was a shortage of manpower for these duties. The LMS had introduced rocking grates to new-build Stanier Black Fives from 1945 and was keen to extend this new technology to the rebuilding of the Royal Scots and Patriots. The first applications were in 1946 on the rebuilt Patriot 45521 *Rhyl* and the rebuilt Scot 6139 *The Welch Regiment*. However, not all locomotives were fitted, as some thirty plus Scots had been rebuilt by

46106 Gordon Highlander (rebuilt September 1949) is seen in 1953 when it was allocated to 8A Edge Hill, before it was fitted with the flat-sided smoke deflectors in July 1954. It was the only rebuilt Class 7 so fitted. The bogie wheels are now the Stanier type, but it retains the Fowler driving wheels. By the time of the new smoke deflector fitment, the engine had been reallocated to Crewe North (in September 1953) and it was fitted with AWS on October 1959. The engine's last depot was Carlisle Upperby, arriving in June 1962 and withdrawn in December 1962 (although it was noted in store in November 1962). However, because of major diesel failures during the severe conditions of the winter of 1962–3, it was reinstated for a few weeks and was seen in London in February 1963 on the 3.30 Upperby to Broad Street meat train, returning the following day with a parcels train. It was also seen at Preston on 16 February 1963 on a parcels train. It went back into store at Kingmoor in March 1963, before being scrapped at Crewe in April 1963. *COLTAS TRUST*

46106 Gordon Highlander *is seen at Cricklewood without a shed plate (but allocated to Carlisle Upperby) or nameplates on 27 January 1963, three weeks after the locomotive had been officially withdrawn during the week ending 8 December 1962 (although it had been stored since October 1962). It was recalled for temporary service during the severe 1962–3 winter.* PETER GROOM

46115 Scots Guardsman (rebuilt August 1947) is seen at Crewe in 1949 sporting its LMS 1946 livery, but with a BR cab side number (applied in January 1949) and its smoke deflectors, which it acquired at rebuilding. This was the only engine to have this livery and smoke deflector combination and when first restored in the 1990s it was given this combination of livery and smoke deflectors.

This shows the original Fowler wheels as fitted to the rebuilt Scots, as well as the single brake shoe and the two centre axle sand pipes. The crank on the right of the picture is for the Smith-Stone speedometer drive, which provides an electric current to a speed indicator in the cab. The small lever above the left-hand sand pipes is the operating arm for the rocker ash pan, to make shed disposal easier and is only fitted to the driver's side of the engine. Not all of the rebuilt Scots had the rocker ash pans, but all the rebuilt Patriots did. The fluted coupling rods can be also seen.

this date, as well as the two rebuilt Jubilees, and some certainly went through to withdrawal without being so equipped.

Mike Bentley, a fireman at Buxton shed, tells of 46129 *Scottish Horse* (9A Longsight) not having rocking bars in 1963 and the engine was missing some of its fire bars. Photographs of this engine taken in the mid-1950s clearly show the rocking gear between the rear driving wheels, so the conclusion is that at least one of the boilers did not have a rocking grate fitted and this boiler would be seen on a number of engines. Certainly, as examples, engines such as 46125 *3rd Carabinier* and 46108 *Seaforth Highlander*, rebuilt in 1943, were not fitted with a rocking grate when rebuilt, but in the 1960s the engines were both seen with a rocking grate.

Footsteps

There was a very subtle change to the footsteps both on the front and rear of the engines; as originally rebuilt only the outer corner of the step was turned up, but later the turn-up was the full width of the step.

Modification of Frames

From 1948, new axle box guides had manganese liners and pin-jointed cross-stays on the coupled wheels. Four cross-stays were also fitted and as well as manganese steel liners to coupled wheel axle boxes.

Platforms and Valances

Early rebuilds had a beaded valance. There were also subtle differences among the classes regarding the valance that hung down from the platform. In the case of the rebuilt Patriots and the two rebuilt Jubilees, the depth of the valance continued to meet the front buffer beam and the rear drag box in a straight line. In the case of the rebuilt Scots, the valance curved down at the ends to the full depth of both the front buffer beam and the rear drag box under the cab. The curving down of the valance at the front followed that of the original Royal Scots, but the curving down to the rear drag box was not seen on them, so the presumption is that the platforms and valances were new when the engines were rebuilt. 46170 *British Legion* had the front of the valance curved in the same way as on the

45529 Stephenson *(rebuilt July 1947) is seen in the goods yard at Chesterfield in August 1948 when celebrations were taking place to celebrate the 100th anniversary of the death of George Stephenson, who died when living in Chesterfield and was buried there. The engine had been rebuilt a year before and had run with a backing plate but no nameplate until it was named on 27 July 1948. At the time of the exhibition, the locomotive was allocated to 1B Camden. It was allocated to Camden a total of five times and seemed to alternate with Crewe North, where it was also allocated five times between June 1948 and January 1961, when the engine moved to Willesden. Its final shed was supposed to have been Annesley, where it was reallocated in November 1963, but it seems to have spent its time stored at Willesden until moving to Crewe Works for scrapping in March 1964. R.J. BUCKLY*

rebuilt Scots, but the rear intersection with the drag box was straight as per the rebuilt Patriots.

Sandboxes on the Footplates

The original Royal Scots had sandboxes below the platform and the first rebuilds continued with this process, but shortly afterwards the distinctive rectangular sandboxes began to appear. This was the result of a report in March 1945, which highlighted that some of the newly rebuilt Scots had been bending coupling rods. This appeared to be partially due to the sanders not applying sand to the rear driving wheels, so to resolve this a double sandbox was to be installed. This was fitted above the platform and had two separate lids, as the box was divided into two, and sand was fed to the front of the rear driving wheels and to the rear of the middle driving wheels. The rebuilt engines that did not have the revised sanding arrangements were soon modified and all new rebuilds of the Royal Scots received the new-style sandboxes.

The rebuilt Patriots had sandboxes below the platform, despite the fact that the original Patriots had sandboxes above the footplate for providing sand to the back of the middle driving wheels. The Patriots also had square-section coupling rods and cylinders 1in (25.4mm) smaller in diameter, which were not affected by the bending of the coupling rods. The two rebuilt Jubilees were similar to the rebuilt Patriots and had sandboxes below the platform, with no sanding to the rear driving wheels.

46170 *British Legion* had similar sandboxes to the rebuilt Jubilees and Patriots.

46162 Queen's Westminster Rifleman (rebuilt January 1948) is seen in 1948 after being renumbered (in the week ending 17 April 1948) and in its superb 1946 LMS livery. It is at a GWR depot, probably when the engine was on the Great Western during the 1948 Locomotive Exchange Trials. 46162 was one of two rebuilt Scots used on various regions; in the case of 46162 it was used on the main line from Paddington to Plymouth.

Top Feed Covers

When originally rebuilt, the top feed cover had a flat surface across the top. However, the Ivatt version had slopes to the top and a small top cover fitted from the 1950s onwards. By the late 1950s, most of the locomotives were fitted with this later style of top feed cover, an exception being 45735 *Comet*, which retained the original cover until withdrawal in October 1964.

Lowered Front Lamp Bracket

Because of issues with overhead wires, most surviving steam locomotives in the 1960s had the top lamp bracket on the smokebox lowered to the 3 o'clock position. WO/E 489 was issued in 1963 to cover this modification. The work was usually carried out at depots, although not all rebuilt 7s running from 1963 onwards had this modification. 45530 *Sir Frank Ree* was photographed at Crewe Works with a chalk mark around the top lamp bracket, with the chalked instruction to move it and a circular chalk mark where the lamp bracket was to be resited. The middle lamp iron on the buffer beam was also usually (but not always) moved to below the revised top iron to avoid

erroneous sightings by signalmen and other railway staff due to the lamps not being in a strict vertical position. Examples that had this modification are as follows:

- 45736 *Phoenix*
- 45526 *Morecambe and Heysham*
- 45530 *Sir Frank Ree*
- 45531 *Sir Frederick Harrison*
- 46115 *Scots Guardsman*
- 46128 *The Lovat Scouts*
- 46152 *The King's Dragoon Guardsman*
- 46160 *Queen Victoria's Rifleman*
- 46166 *London Rifle Brigade*.

Strangely, 46140 *The King's Royal Rifle Corps*, which lasted as long as until 1965, did not receive this modification.

Telltale Devices

These were devices that would give off an extremely strong smell if the inside bearings had run hot and would allow the crew to bring the locomotive to a halt. Job No. NWO/R 1013 of 9 May 1950 was the authorization for the fitting of telltale devices. They were fitted to all the Jubilees, rebuilt Scots, rebuilt Patriots, Princess Royals, Princess Coronations and the original Patriots.

AWS and Speedometer Fitment

Both the LMS and BR had experimented with automated warning systems that would give an audible warning in the cab of a locomotive that a signal was set to danger, which would be particularly useful in foggy conditions. The dreadful accident at Harrow and Wealdstone in 1952 when 112 people died and eighty-eight were injured was caused by a driver passing one signal at caution and two at danger. A report concluded that the accident could have been averted if some form of automated warning had been implemented. BR was already developing such a system, which would automatically apply the brake if a signal was passed at danger, unless the driver cancelled it, and the implementation of its Automatic Warning System (AWS) was accelerated following the accident. The equipment needed to be placed on the tracks (AWS inductors) and a mechanism placed in the cab to give an audible warning if the driver had missed the signals due to foggy conditions and also to apply the brakes automatically. So in 1957 authorization was given to fit AWS to several classes of locos when they were next at Works. This was Job 5797 (Works Order E4983).

Fitting was carried out between February 1959 and September 1960 and the visual changes were as follows for all the rebuilds:

- a contact shoe attached to the front of the bogie
- a 'bash plate' at the bottom of the front buffer beam to prevent the front coupling from swinging into contact with the AWS contact shoe
- a large air reservoir on the platform just in front of the cab front on the fireman's side
- a small air reservoir on the platform on the driver's side
- a conduit clipped to the valance below the platform on the driver's side, curving down into the cab side.

For 46170 *British Legion* the battery box was located on top of the platform just in front of the cab on the fireman's side.

Speedometer Drives

The first two rebuilt Scots, 6103 *Royal Scots Fusilier* and 46109 *Royal Engineer*, were fitted with the BTH speedometer when rebuilt in 1943 and 6170 *British Legion*, already rebuilt, was also fitted with a BTH speedometer. However, all were removed in 1943–4 due to the shortage of spare parts because of wartime conditions.

In 1959, Job No. 5810 (Works Order E5173) authorized the fitting of Smith-Stone speedometers and this was carried out from April 1960 up to May 1962. The work was generally done when the engine next visited Works, but was also dependent upon the availability from the supplier of the necessary components.

The work was cancelled in 1964, but by this time many of the rebuilt Class 7s had been withdrawn and Works visits had virtually stopped, so three rebuilt Scots never received either AWS or speedometers. One rebuilt Patriot, 45514 *Holyhead*, entered Works in May 1961 and would have been fitted with a speedometer, but the engine was withdrawn when it

5690 Leander. *This photo shows the driving wheels and speedometer drive on a Stanier Jubilee and also shows the early style of Stanier wheels, with the webbing on the spokes adjacent to the crankpin. This style of wheel was fitted to several of the rebuilt Patriots, with some finding their way on to a number of rebuilt Scots. The Smith-Stone speedometer was the standard fitment for passenger locomotives in the BR period.*

was stripped down and major problems were found with the frames.

The Engine History Cards do not always show whether either of these items has been fitted (a handful of locos only), or the date. The table in the Appendices summarizes the photographic analysis and the Engine History Cards. All dates are Week Ending; where no date is given the evidence for fitting is photographic. If the comments have SCR (Scottish Region), NER (North Eastern Region) or ER (Eastern Region) it indicates that the engine was transferred to that region and the EHC is therefore likely to be incomplete.

Tenders

As a general rule, the tenders fitted to the classes were as follows:

• Rebuilt Patriots: welded 4,000gal (18,184ltr) tenders. As the original Patriots had the 3,500gal (15,911ltr) Fowler tenders, when they were rebuilt they needed new Stanier 4,000gal tenders and at this point the standard tender was the welded version.
• Rebuilt Scots: riveted 4,000gal (18,184ltr) tenders, as these were the tenders that the original Scots had when they were rebuilt.
• Rebuilt Jubilees: 45735 *Comet* and 45736 *Phoenix*

6115 Scots Guardsman *(rebuilt August 1947). This shows the tender front for the 4,000gal (18,184ltr) riveted tender fitted to 46115 taken at Tysley in the 1980s when the engine was stripped down. The platform immediately below the coal shute in the tender is higher on the rebuilt Class 7s than for other Stanier 4-6-0 locomotives, as the boiler was pitched higher, which meant that the similar tenders from the Jubilees could not be fitted behind a rebuilt Class 7.*

were both fitted with a 4,000gal (18,184ltr) tender, as these were the tenders carried when the engines were rebuilt.

There was, as usual with LMS engines, many tender swaps, as the LMS and later BR had fewer tenders than locomotives, so when a locomotive was in Works it would take the next available tender, which need not necessarily be the tender it entered Works with. However, tender swapping among the rebuilt Class 7s was more limited due to the different height of the fronts

46155 The Lancer *(rebuilt August 1950) is seen at Nottingham Victoria on 19 September 1964 waiting for a Locomotive Club of Great Britain special to arrive, whereupon the rebuilt Scot would take over and pull the train to London. Its home depot of Crewe North has done an excellent job of cleaning the engine and fitting either replica nameplates or the original plates reinstated for the occasion. The engine has its newly acquired yellow warning stripe fitted (applied in late August), which was applied to all the surviving rebuilt Class 7s from that date. The engine was reallocated a few days later to 12A Carlisle Kingmoor, where it remained in service for a few more months until withdrawal in November 1964. The engine remained in store at Carlisle Kingmoor between November 1964 and January 1965, before being scrapped at West of England Ship Breaking, Troon.* MALCOLM CASTELDINE

46155 The Lancer (rebuilt August 1950) shows off its Stanier 4,000gal (18,184ltr) riveted tender, which it had acquired back in 1936 and retained after rebuilding. The majority of the rebuilt Scots ran with the riveted version of the Stanier tender, but a few ran with the welded version. The engine is on shed at Crewe North depot, to which it was allocated on seven separate occasions between rebuilding in 1950 and withdrawal in November 1964. The photo is taken sometime in 1963–4, as the nameplate has been removed, but the yellow warning stripe applied in August 1964 has not yet been painted on. The AWS air tank is visible (fitted July 1959). The engine's final allocation was to Carlisle Kingmoor in September 1964, until withdrawal in November 1964. It was scrapped at West of Scotland Shipbreaking, Troon, in February 1965.

of the tender and the coal plate, which was unique to these engines, so it would not, for example, be possible for a Jubilee tender to be fitted to a rebuilt Class 7.

Some engines managed to keep the same tender throughout their time after rebuilding, such as 45522 *Prestatyn*, but other engines had a number of different tenders, such as 45523 *Bangor*, which had three separate tenders between October 1948 and withdrawal in January 1964. Listed below are the tender changes known to the author, based on photographic evidence and confirmation of the tender numbers listed on the Engine History Cards.

• 45531 *Sir Frederick Harrison* was originally fitted with a 4,000gal (18,184ltr) welded tender and then was fitted with a 4,000gal riveted tender (No.9370) from 1956 until withdrawal in 1965.

• 45534 *E. Tootal Broadhurst* was originally fitted with a 4,000gal (18,184ltr) welded tender (No.9765) and was then fitted with a riveted tender (No.9037) in 1953. It reverted to a welded tender (No.9781) in September1960.

• 45536 *Private W. Wood, V.C.* had a welded tender (No.9781) from rebuilding, then a riveted tender (No.9037) from July 1960 until withdrawal.

• 46104 *Scottish Borderer* was fitted with a 3,500gal (15,911ltr) riveted tender (No.4620) in 1937 and

kept this tender following rebuilding in 1946 up until October 1955, when it acquired another 3,500gal riveted tender (4648). It kept this up to March 1961, when, following a Works visit, it acquired a 4,000gal (18,184ltr) riveted tender (No.9041).

• 46123 *Royal Irish Fusilier* was fitted with a 4,000gal (18,184ltr) welded tender April 1953 until November 1957.

• 46132 *The King's Regiment Liverpool* was fitted with a 4,000gal (18,184ltr) welded tender from July 1952 to November 1952.

• 46142 *The York and Lancaster Regiment* was fitted with a 4,000gal (18,184ltr) welded tender (No.9765) from November 1957 onwards and another welded tender (No.9779) from 1959 until withdrawal in January 1964.

• 46162 *Queen's Westminster Rifleman* was fitted with a 4,000gal (18,184ltr) riveted tender from 3 July 1952 (No.9041) until 21 March 1961, when it swapped tenders with 46104 Scottish Borderer and acquired a 3,500gal (15,911ltr) riveted tender (No.4648), which it kept until withdrawal.

• 46165 *The Ranger* (12th London Regiment) was fitted with a welded tender (No.10379) between August 1960 and withdrawal in November 1964.

6104 Scottish Borderer (rebuilt March 1946) is seen at Polmadie in the 1940s and shows off its LMS livery of black and LMS lettering. The tender is unusually the 3,500gal (15,911ltr) riveted tender fitted before conversion and which it retained until October 1955. The top feed is of the original type with the flat top and the engine is yet to be fitted with smoke deflectors (fitted August 1951). The engine went to Polmadie in 1942 before conversion and remained at the depot until transferred to 67A Corkerhill in October 1962 for a few months before withdrawal in December 1962.
BLENCOWE COLLECTION

45522 Prestatyn (rebuilt February 1949) sits at its home depot of Longsight on 22 September 1963, where it had arrived from the other side of Manchester, Newton Heath, in June 1963. The engine displays the typical care-worn condition of the early 1960s, although when later used by Buxton shed they gave it a good clean for its use on the passenger train to Manchester Central. The welded tender fitted when rebuilt is still attached to the engine. RAILONLINE

46142 The York and Lancaster Regiment (rebuilt February 1951) is seen with its welded tender, one of the few rebuilt Scots to have the welded version. Tender No.9779 was acquired at a Heavy General overhaul in August 1959. The engine has a 14B Kentish Town shed plate. It was allocated there in September 1959, when a number of rebuilt Class 7s were transferred to the Midland Main Line. It remained at Kentish Town until transfer away to Newton Heath in September 1961. The nameplate has a red painted background and the regimental badge is missing. The engine has Stanier-style leading driving wheels and balance weights. JOHN FOZZARD

46165 The Ranger 12th London Regiment (rebuilt June 1952). How are the mighty fallen, with 46165 shunting the yard at Llanrwst & Trefriw on the branch from Llandudno Junction to Blaenau Festiniog in July 1963. The engine at the time was allocated to Crewe North (arriving in June 1963), but had previously been allocated to Llandudno Junction (arriving in September 1962), which would be more appropriate to working down the branch to Blaenau. Why it had been sent down the branch will remain a mystery, as the biggest engines usually seen there were the Ivatt Class 2 2-6-2 tanks. The welded tender that the engine acquired in August 1960 is clearly visible, as are the AWS (June 1959) and speedometer (August 1960). The engine moved from Crewe North in February 1964 to Annesley and was the last rebuilt Scot working at the depot when it was withdrawn from there in November 1964. I. TRAVERS

45534 E. Tootal Broadhurst (rebuilt December 1948) is seen at Bletchley in April 1960 with the fireman trimming coal in the 4,000gal (18,184ltr) riveted tender. The rebuilt Patriots usually ran with welded tenders as these were provided from the rebuilding date, but a number of them acquired riveted tenders during their lives. 45534 had a welded tender when rebuilt and acquired its riveted tender in April 1953, then reverted to a welded tender in September 1960. The engine acquired AWS in April 1959 and a speedometer in September 1960. At the time of the photograph, the engine was allocated to Llandudno Junction, where it arrived in November 1959 and departed for Longsight in June 1960. RAIL BRM

45534 E. Tootal Broadhurst (rebuilt December 1948) is seen at Patricroft shed on 27 October 1961, whilst it was allocated to Llandudno Junction. Patricroft did not have any rebuilt Class 7s allocated, but they were regular visitors as the Holyhead and Llandudno engines would be serviced at Patricroft after working North Wales–Manchester Victoria and return services. The engine remained at Llandudno, before moving to Crewe North in June 1963, where it stayed until withdrawal in May 1954, followed by disposal at Crewe Works in June 1964. The engine is fitted with AWS (April 1959) and a speedometer drive (September 1960). It has a welded tender, this being the third tender fitted to the engine and was retained until disposal. COLTAS TRUST

45527 Southport (rebuilt September 1948) is seen at Crewe North in what looks like ex-Works condition, but with an Edge Hill shed plate, where it had been allocated from December 1933 until it was transferred to Bushbury in March 1961, before moving to Holyhead in May 1961. Allocations to Holyhead and Willesden followed, before arriving at Carlisle Kingmoor in September 1963. It remained at Carlisle, moving between Upperby and Kingmoor, before being withdrawn in December 1964 and scrapped at West of Scotland Shipbreaking, Troon, in February 1965. The engine has a welded tender and the AWS tank can be seen (March 1959) and the bogie wheels are Fowler style. The name proudly displays the crest of the Borough of Southport above the nameplate. The axles have holes through the middle, which could be seen on a number of rebuilds.

46143 The South Staffordshire Regiment (rebuilt June 1949) is seen sporting its newly acquired BR black, lined out in red and grey based on the old LNWR livery and with 'British Railways' on the tender. The smokebox number plate is the standard style. The engine has been fitted with the later revised top feed and the leading coupling rods have the revised circular bush.

CHAPTER THREE

LIVERIES, NAMES AND NAMEPLATES

Liveries

The liveries applied to the rebuilt Class 7s were varied during the initial years of rebuilding, as the first rebuilds occurred during World War II, when the classic LMS Crimson Lake red livery with gold lettering with shading had to give way to an economy black with a variety of lettering. The only engine to carry the classic LMS Crimson Lake was 6170 *British Legion*, as it had been rebuilt in 1935, so it carried this livery until the early 1940s.

With the ending of the war, the LMS unveiled its new passenger livery in 1946. This was black relieved by a maroon on the edge of the platform, as well as a maroon strip 2in (51mm) wide on the cab edge and tender sides; the maroon was edged with a straw line. This livery was short-lived, as the formation of British Railways in January 1948 led to a period of change, with the rebuilt Class 7s initially being given a black livery with the new BR mixed-traffic lining of grey/cream and red in the style of the LNWR with 'British Railways' in full on the tender sides.

As part of the process to decide on the new BR passenger livery, a number of engines were painted in experimental liveries and this included three rebuilt Class 7s, with 45540 *Sir Robert Turnbull* and 46139 *The Welch Regi-*

45531 Sir Frederick Harrison *(rebuilt December 1947) is seen at Derby in May 1948 after being repainted in an experimental livery. Two other rebuilt Class 7s were painted in this experimental light green livery, 45540* Sir Robert Turnbull *and 46139* The Welch Regiment. *Fortunately, this rather bilious shade of green was not continued with.* COLOUR-RAIL

ment and 45531 *Sir Frederick Harrison* both receiving a livery of 'experimental light green', which appeared close to the apple green formerly applied to LNER passenger engines. In the case of 45540, it was applied at Derby Works rather than Crewe. This livery did not sit well on the engines and was not continued; both engines had the livery covered up with BR green a year later.

In 1949, the livery issue was finally decided, when BR decided to paint the rebuilt Class 7s in the new BR passenger livery of Dark Bronze Green lined out GWR style with orange and black lining. Some sources say the green was Brunswick Green, but the correct term was Dark Bronze Green. As the rebuilt Class 7s were seen in Works almost on an annual basis, it did not take long for the locomotives to acquire this livery, which they then carried until withdrawal in the 1960s. Of course, some engines were rebuilt after the introduction of the new green livery and therefore only ever carried the BR livery until withdrawal.

The transition to the BR livery also included some interesting variations in both front smokebox number plates and to the cab side lettering, with some engines having an 'M' before the number prior to them being given full BR numbers; in some cases, the BR number

was applied as well as an 'M'. The lettering style on the front number plates also varied, with some BR numbers being cast using a scroll and serif lettering before being replaced by the standard BR lettering style using sans serif letters. The only other livery change was to the BR emblem on the tender from 1957 onwards.

LMS 1942 to 1946

The first two rebuilt Jubilees, 5735 *Comet* and 5736 *Phoenix*, were both out-shopped in 1942 in black with shaded yellow lettering, as the LMS had decided that in 1940 any locomotive requiring repainting should be in black. Given the wartime conditions and the complete lack of cleaning, it was a mystery what colour the engines were painted or whether the lettering was shaded or not. The running number was 12in (305mm) high and was placed centrally on the cab side and the power classification '6P' was below the cab windows. This placed the cab side number below the 'LMS' on the tender, but in 1945 the cab side number was raised in line with the 'LMS' and the power classification dropped below the cab side number. Some

6117 Welsh Guardsman *(rebuilt December 1943) is seen at Crewe North shed shortly after rebuilding. The leading set of driving wheels has webbing behind some of the spokes, but does not have a bevel rim, so these wheels are probably from a Patriot. The livery is black, but the cab side lettering is not clear and the 'LMS' is not seen on the tender, so the date is between the end of 1943 and May 1948, when it was renumbered.*

6115 Scots Guardsman *is seen after its restoration into the LMS 1946 livery at Dinting in the 1970s. This engine was rebuilt in August 1947 and was the only rebuilt Scot to carry this livery and to have smoke deflectors.* Scots Guardsman *has now been overhauled and repainted in the standard BR passenger green livery and is occasionally seen on the main line.*

of the cab side and tender lettering was shaded and some not, but given the general filthy state of the engines it is not always clear.

LMS Post-1946

From the middle of 1946, any newly converted engines, as well as some of the already converted ones, received the new LMS passenger livery. This livery was a bit of a come-down from the pre-war LMS red, as the base colour was black with a maroon edge to the plat-form and a straw line at the top and bottom of the maroon panel. The cab side and tender sides also had a maroon edge, again lined out in straw. The lettering style was also different from the previous LMS letter-ing, being more like a 'block' style of lettering. Engines known to have received this livery include: 5514, 5521, 45526, 5529, 45530, 5531, 5532, 6103, 6108, 6112, 6117, 6120, 6121, 6124, 6127, 6131, 6132, 6133, 6135, 6138, 6145, 6146, 6147, 6149 and 6166.

6118 Royal Welch Fusilier *(rebuilt December 1946) stands at Crewe circa 1947, displaying the LMS 1947 livery. It is fitted with a full set of the Fowler driving and bogie wheels and has the early style of top feed cover. The nameplate has a crest underneath of a Welsh dragon, although this is not the regimental badge.* ROGER CARPENTER'S COLLECTION

British Railways 1948 Onwards

A number of engines were seen with the LMS 1946 livery, including 'LMS' on the tender, but with a BR cab side number combined with 1946 style lettering; examples include 46132. Some engines carried the LMS 1946 livery, but had standard BR cab side numbers applied; examples include 45526 and 46115. A number of engines that carried the LMS 1946 livery had 'British Railways' on the tender and were renumbered; engines seen with this livery include 45512, 45514, 45526, 45532, 46111, 46127, 46128, 46132 and 6138 *The London Irish Rifleman*. In some cases, the BR number was in the same numbering transfer style of the 1946 livery, but with smaller numbers, although these numbers were larger than the later BR cab side numbers; examples include 45514, 45532, 46106, 46117, 46120 and 46170. 46154 was seen in 1948 in 1946 LMS livery with 6154 on the cab side, but with 'British Railways' on the tender.

Some engines ended up for a short period of time with 'hybrid' liveries, so 46116 , 46145, 46161, 46162 and 46166 had BR numbers on the cab side in LMS 1946 style, but the tender had the pre-war LMS style on the tender, along with the power classification above the cab side number. The cab side number was quite high on the cab side.

The initial 'official' BR livery was the mixed-traffic livery of black, lined out LNWR style with grey, cream and red. This had 'British Railways' on the tender. Locomotives known to have carried this livery include: 45522, 45523, 45525, 45527, 45529, 45534, 45535, 45536, 45735, 45636, 45735, 45736, 46112, 46121, 46122, 46123, 46127, 46128, 46139, 46143, 46144, 46150, 46160, 46167 and 46169.

The final BR colour of Dark Bronze Green was applied with orange and black lining, similar to that used on the Great Western passenger locomotives.

46132 The King's Regiment Liverpool (rebuilt November 1943). The transition from LMS to BR produced some interesting livery variations and here the base livery is LMS 1946. However, a BR number plate is in the scroll and serif style that did not last long before being replaced with a different lettering style. The BR cab side number is in a style similar to the LMS 1946 lettering and was replaced by the standard BR lettering.

6138 The London Irish Rifleman (rebuilt in June 1944) received Works attention in 1946, where it is assumed it acquired its 1946 livery. The photograph of this early hybrid LMS/BR livery was taken about 1948, as the engine was renumbered 46138 in January 1949. The locomotive is in its 1946 LMS livery, but an M has been added to the cab side number and the smokebox number plate as well as British Railways added to the tender. The top feed is the original type with a flat top and the bogie wheels are the original Fowler type, as are the driving wheels. The front coupling rod has the revised circular bush, which replaced the rectangular split brasses fitted to the original engines, but which were retained for a short period on some of the rebuilds.

46139 The Welch Regiment (rebuilt November 1946) is seen after being repainted in the experimental 'apple green' in 1948, when BR was wanting to decide on a passenger livery and a number of engines were painted in different liveries. Another two rebuilt Class 7s were also painted in this livery. The engine has the split brasses to the leading coupling rods and the crest above the name is missing.

46127 Old Contemptibles *(rebuilt August 1944) is seen in 1949 in the LMS 1946 livery, but with a BR number on the cab side (although the lettering style is similar to the LMS 1946 style) and the short-lived serif-style front number plate. The view shows much of the details on top of the platform and also the very 'non-standard' nameplate that was given to this engine (originally named 6127 Novelty) in November 1936.*

45526 Morecambe and Heysham *(rebuilt February 1947) is seen circa 1949 and shows it hybrid livery, which is LMS 1946, but with a BR number on the cab side and smokebox number plate. The tender is the welded one it acquired when rebuilt and kept until it was withdrawn in October 1964. The top feed is the early type with a flat top. The feed pipe to the top feed is more visible under the footplate than on the rebuilt Scots.*

45535 Sir Herbert Walker, K.C.B. (rebuilt September 1948) is seen at Crewe shortly after rebuilding and is resplendent in its BR livery based on the LNWR passenger livery. The smokebox number plate has the initial style of lettering, which was later revised. The engine was to receive its BR passenger green livery sometime around 1949. The shed plate is 5A Crewe North, to which it was allocated from its rebuilding until a transfer to Camden in September 1950. The engine returned to Crewe North on three separate occasions, before moving to Edge Hill in June 1954, where it remained until its final move to Carlisle Kingmoor in October 1962. Smoke deflectors remain to be fitted (January 1952), as has the AWS (January 1959) and speedometer drive (June 1960). The engine has a welded tender, which it received upon rebuilding and retained until withdrawal.

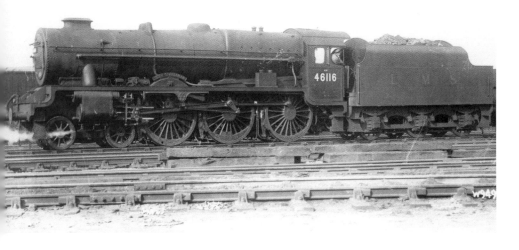

46116 Irish Guardsman (rebuilt May 1944) displays one of the many transitional liveries during the early BR period. In this case, the tender is lettered in pre-war LMS style, but the new BR number applied in September 1948 is in the style of the LMS 1946 lettering, which is larger than the standard BR lettering applied later. The lettering is also higher up the cab side than the later BR numbers. The hole in the front driving wheel seems very close to the edge of the wheel. REAL PHOTOS

45526 Morecambe and Heysham (rebuilt February 1947) is seen at Crewe in the early 1960s, with the engine fitted with AWS (in November 1959) and overhead warning flashes. As with most of the rebuilt Patriots, the wheels are all of the Stanier type with bevelled rims and the driving-wheel axles have holes through them. The coupling rods are of plain section, unlike the rebuilt Royal Scots, which were fluted. The engine was allocated to Leeds Holbeck (four weeks), Edge Hill (two years) and Bushbury for eight months, before becoming a long-term resident of Carlisle Upperby, where it was allocated in June 1950 and remained until October 1964. It was scrapped in February 1965.

45526 Morecambe and Heysham (rebuilt February 1947) is seen in an official photograph at some point in the mid-1950s. The welded tender has the pre-1957 tender emblem and the shed code on the smokebox is 12A, which at the time was Carlisle Upperby (it changed to 12B in February 1958). The engine received AWS in November 1959 and a speedometer drive in February 1961.

Nameplates

The naming policy for the rebuilt Class 7s was a bit of a mess and certainly lacked the consistency of GWR naming policy. When the Royal Scots were originally built, a number of the locomotives were named after British Army Regiments and another batch was named after early locomotives of the Liverpool and Manchester Railway (such as 6127 *Novelty* and 6128 *Meteor*), but in 1935–6 all of these locomotives were renamed with more British Army regimental names.

This meant that the rebuilt Scots ended up with a consistent set of names, being in the main named after Regiments in the British Army (or individual soldiers from British Regiments), although there were a few exceptions (46149 *The Royal Air Force*, 46168 *The Girl Guide*, 46169 *The Boy Scout* and 46170 *The British Legion*).

The first twenty-five locomotives of the rebuilt Scots (with three exceptions, 46102 *Black Watch*, 46121 *Highland Light Infantry* and 46124 *London Scottish*) were named after individual solders, rather than the Regiment (such as 46105 *Cameron Highlander* and 46122 *Royal Ulster Rifleman*). A number of other locomotives were also named after individual solders rather than the full Regiment, such as 46150 *The Life Guardsman*, 46151 *The Royal Horse Guardsman*, 46152 *The King's Dragoons Guardsman*, 46154 *The Hussar* and 45155 *The Lancer*.

The naming policy for the original Patriots was also a little bit random, with a mixture of ex-LNWR directors (45530 *Sir Frank Ree*), Victoria Cross winners (45536 *Private W. Wood, V.C.*), town boroughs that were served by the LMS (such as 45527 *Southport*) and names from very early locomotives (such as 45545 *Planet*). As some of the original Patriots were unnamed when they were rebuilt, it meant that a new name would have to be allocated as it would appear that the rebuilds should always be named, so when 5529 was rebuilt in July 1947 it had a backing plate for a name but was not named *Stephenson* until July 1948 on the 100th anniversary of George Stephenson's death. 5528 was similarly a rebuild of an unnamed member of the Patriots and became 45528 REME. Having been rebuilt in 1947, it remained unnamed as a rebuild until August, or even October,

1959, so it ran longer unnamed as a rebuild than it did with a name.

When locos were in Works it was possible for nameplates to be placed back on the wrong engine, particularly if the engines were next to each other being rebuilt, which would require the whole engine to be dismantled. For example, on 3 August 1943, 6125 *3rd Carabinier* left Works with the nameplate from 6108 *Seaforth Highlander* and two weeks later 6108 left with the nameplate from *3rd Carabinier*. An eagle-eyed member of the RCTS spotted the mistake and alerted the railway authorities and the error was corrected on 28 August 1943.

In some cases, locomotives officially swapped name and number for a short period; as an example, in September 1955 46132 *The King's Regiment Liverpool* was repainted and renamed 46121 *Highland Light Infantry*, as it was required to haul a train containing the Regiment and the 'real' 46121 was in Works for repair. The 'imposter' 46121 returned to Crewe the following week to be repainted back to its original name and number.

46109 Royal Engineer (rebuilt July 1943). This shows a standard single-line regimental name with a crest above the name. The engine was named seven months after being built in September 1927. The engine, after rebuilding, was allocated to Camden for a few weeks, before arriving at Leeds Holbeck in August 1943. It remained at the depot until September 1961, when it transferred to Low Moor Bradford, before returning to Holbeck in June 1962 and being withdrawn in December 1962.

46141 The North Staffordshire Regiment *(rebuilt October 1950). This is an example of a name with a double line and a crest above the nameplate. The original Royal Scot had been named* Caledonian, *but was renamed* The North Staffordshire Regiment *in April 1936.*

46130 The West Yorkshire Regiment *(rebuilt December 1949) displaying its double-line nameplate and backing for its regimental crest. There were a number of instances where a locomotive could be seen with one or more of its crests missing from the backing plate. The engine had been named* Liverpool *after an early locomotive, but was renamed in June 1935. From November 1959, 46130 was allocated to 55A Leeds Holbeck and withdrawn from that depot in December 1962 (with a brief sojourn at Low Moor Bradford from September 1961 to June 1963).*

46118 Royal Welch Fusilier. *An example of one of the few rebuilt Scots that had a crest below the nameplate, as the majority had the crest above the name. The name is actually of an individual soldier; the regimental name would be 'Welch Fusiliers' and the badge below the plate is not the regimental cap badge. The crest below the name is also unusual, in having the top portion cut off by the name above. As one of the oldest Regiments in the Army (being raised in 1689 as the 23rd of Foot), it used the archaic spelling of 'Welch' instead of 'Welsh'. The engine spent six years at Crewe North shed, before being moved a number of times in the space of two years from 1960, when it moved to 16A Nottingham (in January 1960), then 21A Saltley (in August 1961) and then Carlisle Upperby in June 1962, remaining at Upperby until withdrawal in June 1964.*

45545 Planet (rebuilt November 1948) is seen on a special train at Coventry station. The opening of the new cathedral in May 1962 generated a huge amount of special trains in 1962–3, originating from all over the country. 45545 Planet displays its Carlisle Upperby shed plate, where it had been allocated from June 1961 and remained at the depot until withdrawal in May 1964. When the locomotive was withdrawn and was stored awaiting scrapping (at Connels, Coatbridge, in September 1964), it was still in excellent external condition.

45545 was named Planet at its conversion in 1948 and the name had previously been carried by an original Royal Scot, 6131, which had been renamed The Royal Warwickshire Regiment in March 1936. Planet's bogie wheels are Fowler and the driving wheels look like the later Fowler type, but with Stanier-style riveted balance weights. The tender is the welded version given to the engine on its conversion and the AWS air tank can be seen on the footplate near the cab.

45530 Sir Frank Ree (rebuilt October 1946). The names applied to the Patriots and rebuilt Patriots were a strange collection with little consistency of names. Sir Frank Ree was an LNWR General Manager and the name had been originally carried by an LNWR Claughton. New plates were cast when Patriot 5902 (later renumbered to 5501) was built, but the plates were transferred to another Patriot (6022, later 5530) when 5902 was renamed St Dunstans in 1937. The plates were certainly still on the engine in July 1963, but had been removed by early 1965.

45527 Southport. Like the regimental names on the rebuilt Scots, which had a crest above or below the nameplate, the rebuilt Patriots named after seaside resorts carried the Borough coat of arms above the nameplate, as illustrated here. Similar coats of arms were carried by 45521 Rhyl, 45522 Prestatyn, 45525 Colwyn Bay and 45526 Morecambe and Heysham. The nameplates for 45514 Holyhead and 45523 Bangor did not have Borough coats of arms above the plates.

45528 REME (rebuilt August 1947). When this engine was rebuilt it was unnamed and unusually remained so until October 1959. The original intention was that a crown be mounted over the name and a drawing was done, but the crown was never made or fitted. The photograph was taken in the mid-1950s and the engine has still to be fitted with AWS (January 1960). *R.K. BLENCOWE*

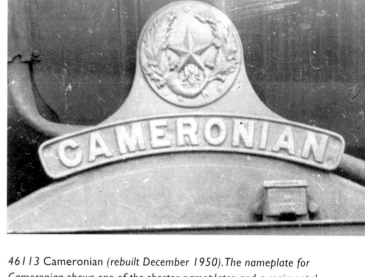

46113 Cameronian (rebuilt December 1950). The nameplate for Cameronian shows one of the shorter nameplates and a regimental crest above the plate. The engine was named in March 1928 and is an example whereby the name represents a single soldier rather than the Regiment, which was The Cameronians (Rifle Brigade). The engine was a long-term resident of Leeds Holbeck after rebuilding, only moving away to 56F Low Moor in November 1961, before moving again to Mirfield in January 1962, then back to Leeds Holbeck in June 1962, before withdrawal in December 1962.

The nameplate for 46162 Queen's Westminster Rifleman is seen on the engine on 18 April 1964. It was unusual for the rebuilt 7s to retain their nameplates, some having been removed as early as 1962. The engine was named in 1932 and shows a two-line nameplate, but without a crest. Whether a loco had a crest or not depended on whether the Regiment concerned was prepared to pay to have them made and fitted to the engine concerned. *KEN TYLER*

45522 Prestatyn (rebuilt February 1949). A small number of Patriots were named after seaside resorts served by the LMS, with groups from the Fylde coast (Southport, Blackpool and so on) and also from the North Wales coast (Rhyl, Colwyn Bay and so on), with Prestatyn shown here. This was an example of a plate that did not carry a crest above it with the Borough coat of arms, as these had to be paid for by the Borough concerned. 45522 Prestatyn retained its nameplate until withdrawal in September 1964.

46121 Highland Light Infantry, City of Glasgow Regiment (rebuilt August 1946). This is an example of one of the rebuilt Scot nameplates that had a single line for the name with a second line with smaller letters underneath. The original Scot nameplate was the shortened version of 'H.L.I.'; the rebuilt Scot was renamed and given the full regimental name on 15 January 1949, when the plates were fitted in Crewe Works. However, the engine was seen without nameplates until 22 January 1949, when the crests above the name were presented by the Lord Provost of Glasgow (as well as a detachment of the Regiment) at a ceremony at Glasgow Central station. Following rebuilding, the engine had a short stay at Longsight, but was transferred to Polmadie in June 1949 (initially on loan), where it remained until withdrawal in December 1962. It was stored first at Polmadie and later at 65C Parkhead, before being scrapped at McWilliams at Shettlestone in May 1964.

6125 Lancashire Witch. A number of the original Royal Scots were named after early locomotives and as part of these names a small plaque was placed under the name with an outline of the original locomotive. The use of plaques for these engines generated an interest in putting regimental plaques over or under the respective Regiment's name. All of the engines named after early locomotives were renamed in 1936 after Regiments of the British Army. 6125 was renamed 3rd Carabinier. The plaque under the name meant that the nameplate was raised above the splasher.

46138 The London Irish Rifleman (rebuilt June 1944) shows another variation to the nameplates, in this case the placing of a small plaque under the nameplate. A number of other locomotives, including 46116 Irish Guardsman, 46168 The Girl Guide and 46169 The Boy Scout, also had these small plaques under the nameplate. L. HANSON

46155 The Lancer (rebuilt August 1950). This illustrates an example of a short single-line nameplate fitted to the rebuilt Scots. The name refers not to a Regiment, but to an individual cavalry soldier and at the time of the engine being named in 1931 there were three different Regiments of Lancers in the British Army. The photo was taken in 1963 and when the engine was photographed in early 1964 at Crewe the nameplate had been removed. However, this was common if the engine was stored serviceable. KEN TYLER

The mounting of the nameplates also revealed lots of small detailed differences, as follows:

- Single-line plates that were mounted directly above the splasher top. Examples include: 46155 *The Lancer*, 46149 *The Middlesex Regiment* and 46159 *The Royal Air Force*.
- Single-line plates that were mounted with a gap between the top of the splasher and the bottom of the nameplate, the gap being taken up by the backing plate to which the nameplate was attached. Examples include 45522 *Prestatyn* and 46129 *Scottish Horse*. Certainly on some of the engines this gap under the plate was based on the original name, such as 6129 *Comet* having a plaque representing the original engine located under the plate. This would require the nameplate to be pushed up higher than the splasher top. When the engine was renamed *Scottish Horse* the single-line plate had a gap between the bottom of the plate and the splasher.
- Single-line plates mounted directly above the splasher, but with a crest above the plate. Examples include: 45527 *Southport* and 46113 *Cameronian*.
- Single-line plates mounted above the splasher, but with a crest below the nameplate. Examples include: 46102 *Black Watch*, 46116 *Irish Guardsman*, 46118 *Royal Welch Fusilier*, 46122 *Royal Ulster Rifleman*, 46123 *Royal Irish Fusilier*, 46158 *The Loyal Regiment*, 46161 *King's Own*, 46168 *The Girl Guide* and 46169 *The Boy Scout*.
- Double-line plates mounted directly above the splasher top with a crest below the plate. Examples include 46120 *Royal Inniskilling Fusilier* and 46138 *London Irish Rifleman*.
- Double-line plates mounted directly above the splasher. Examples include: 46143 *The South Staffordshire Regiment*.
- Double-line plates mounted directly above the splasher with a crest above the plate. Examples include: 46148 *The Manchester Regiment*.
- A plate which did not conform to the normal style was that fitted to 46165 *The Ranger* (12th London Regiment), which had 12th London Regiment underneath the name.
- Two other nameplates that also did not conform to the standard type were those fitted to 46127 *Old Contemptibles* and 46170 *British Legion*, which were effectively badges rather than a nameplate. The *Old Contemptibles* refers to soldiers of the regular British Army who served at the start of the First World War (1914), before conscription started, and were awarded the Mons Star in 1917. After the end of the war The Old Contemptibles Association was formed and the nameplate of the engine was a copy of the lapel badge of the Association. The engine was renamed with the badge-style plate in 28 November 1936, having run with a conventional-style single-line nameplate, *The Old Contemptible*.

46165 The Ranger (12th London Regiment). One of the nameplates from the engine was recently sold at auction and an opportunity was taken to photograph it. The plate did not conform to the more standard style of plate that was placed on the rebuilt Scots. The name relates to a London Territorial Regiment. TONY SHEFFIELD

46170 British Legion (rebuilt October 1935). When 6399 'Fury' was rebuilt it was also renamed British Legion. The background to the nameplate was usually painted red. After the withdrawal of 45500 Patriot (which was designated as the memorial engine) in March 1961, British Legion was taken to Rugby shed on 11 November 1961 and again on 11 November 1962, where it formed the centre piece of the Remembrance ceremony at the shed. For the event in 1962, the engine was removed from storage at Llandudno Junction shed specifically for the ceremony, before the engine was withdrawn later in November 1962.

46127 Old Contemptibles (rebuilt August 1944), whose nameplate was not the usual shape. The badge name was derived from the Mons Star (also known as the 1914 Star) issued to all members of the British Expeditionary Force (BEF) who had served in France between 5 August and 22 November 1914. The badge was a copy of the Old Contemptibles Association lapel badge, the only difference being the deletion of the word 'Association' from the lower part of the circle, which on the locomotive name was left blank. The engine had originally carried a nameplate similar in style to the other Royal Scots, but was changed to the badge style at a ceremony at Euston on 28 November 1936.

Plaques and Embellishments

With the naming of the Royal Scots when first built, Sir Henry Fowler early in 1928 conceived the idea of placing a small plaque under the name of the twenty-six locomotives named after pioneer engines (such as Novelty). These plaques provided excellent publicity and some of the Regiments which had Royal Scots named after them put forwards the suggestion that regimental badges be displayed on the engines and the Regiments would pay. It was not clear which engine was the first to receive a plaque, but it is likely to have been 6118 *Royal Welch Fusilier* in the spring of 1929.

Subsequently, many of the Regiments raised money to allow crests to be fitted to 'their' engine. In some cases, the Regiment itself did not fund the crest, but would appeal to members (or ex-members) for funds, as in the case of London Rifle Brigade, who raised £20 on appeal to pay for the crest. Most of the plaques were fitted above, but a small number had a plaque fitted below the name despite there being less space.

Locomotives fitted with plaques above the name include: 46103, 46107, 46109, 46112, 46113, 46124, 46125, 46126, 46128, 46129, 46130, 46131, 46132, 46133, 46134, 46135, 46136, 46137, 46138, 46139, 46140, 46141, 46144 to 46149 and 46166.

Crests above the names were also applied to some of the rebuilt Patriots, usually the civic arms for those named after boroughs (such as Southport, Morecambe and Heysham, Colwyn Bay), although, like the rebuilt Scots, not all the locomotives named after boroughs had the civic crests above the name.

When rebuilt, Patriot 45528 *REME* was finally named in 1959. It was intended that a crown be placed over the name, but although a drawing was done, the crown was never made nor fitted. One of the two nameplates has in preservation been mounted with the intended crown above the name.

46112 *Sherwood Forester* had a rectangular plate under the nameplate on both sides which listed the occasions that the locomotive had pulled trains filled

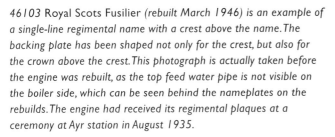

46103 Royal Scots Fusilier (rebuilt March 1946) is an example of a single-line regimental name with a crest above the name. The backing plate has been shaped not only for the crest, but also for the crown above the crest. This photograph is actually taken before the engine was rebuilt, as the top feed water pipe is not visible on the boiler side, which can be seen behind the nameplates on the rebuilds. The engine had received its regimental plaques at a ceremony at Ayr station in August 1935.

46129 The Scottish Horse (rebuilt December 1944). This shows the variation among rebuilt Scot nameplates for a single line and crest above, but in this case it shows the plate is raised above the splasher, when most single-line plates were flush with the splasher top. The previous name had been Comet *and because a plaque had been mounted under the nameplate, the plate had to be raised above the splasher top and this feature was retained when the engine was renamed.* MIKE BENTLEY

with the Regiment. The plates were formally unveiled by Brigadier P.N. White at a ceremony at Nottingham Midland station on 18 April 1948 and the locomotive was crewed by two ex-Sherwood Forester Regiment members, Driver Croll and Fireman Follon, who both proudly wore their military medals on the footplate. The details on the plaque are as follows:

• 1st Battalion, Willesden to Lichfield, 15 November 1932
• 5th Battalion , Crewe to Holyhead, 30 July 1933
• 2nd Battalion, Liverpool to Derby, 8 November 1947
• 5th Battalion, Rotherham to Derby, 18 June 1960
• 8th Battalion, Nottingham to London, 30 July 60
• 5th/8th Battalion, Nottingham to London, 24 June 1961
• 5th/8th Battalion, London to Derby, 8 July 1961.

It would be interesting to know who was responsible for keeping this plate updated and when and where it was done.

Some nameplates had the background painted red instead of the more usual black. Known examples include:

• 45532 *Illustrious* in February 1960, when allocated to Nottingham
• 46100 *Royal Scot*; the name only had a red background in 1961, with the commemorative plate below having a black background
• 46103 *Royal Scots Fusilier*, in the early 1960s
• 46132 *The King's Regiment Liverpool*, circa 1961
• 46137 *The Prince of Wales's Volunteers* (South Lancashire), circa April 1962 until withdrawal
• 46142 *The York and Lancaster Regiment*, circa 1963
• 46157 *The Royal Artilleryman*, circa 1961
• 46170 *British Legion*.

The LMS 1946 livery also listed that the nameplate background colour should be maroon, but it has not been possible to verify if any engines were painted in this way.

Many locomotives, whilst still retaining their name-

46112 Sherwood Forester (rebuilt September 1943). As well as its nameplates and regimental crests, the locomotive also carried plates under the name describing the occasions when the Regiment was hauled by the locomotive. It was common for Regiments to request to be hauled by a locomotive bearing the regimental name.

46100 Royal Scot (rebuilt June 1950) displays the nameplate with the plate below which commemorates the journey that the locomotive made in 1933 when it toured the USA. The commemorative plate was modified after rebuilding, with the words 'prior to rebuilding' at the top of the commemorative part. When in service, the engine was observed with the nameplate 'Royal Scot' with a red background colour and the background to the commemorative plate in black.

46137 The Prince of Wales's Volunteers (South Lancashire) (rebuilt March 1955). The usual backing colour on the nameplates was black, but a small number of engines had the background painted in red, as seen here. The photo was taken at Crewe Works, where the engine had arrived for scrapping in April 1963. It was scrapped the following month, having been withdrawn from Carlisle Upperby in October 1962, with the engine surviving six months in storage with the nameplates and crests remaining in place. RAILONLINE

plates, could be seen with missing plaques on one or both sides. Examples include:

- 46130 *The West Yorkshire Regiment* was seen on 18 July 1962 with at least one crest missing
- 46140 *The King's Royal Rifle Corps* was seen in 1960 following Works attention with one of the crests missing and the space where the crest should have been painted green
- 46142 *The York and Lancaster Regiment* was seen in 1959 with the crest on the left-hand side missing.

Nameplates began to be removed from the rebuilt Class 7s in late 1961, with the first recorded example being 46112 *Sherwood Forester*, which had its plates removed a few weeks after a Class 45 (Peak) D100 was given the name *Sherwood Forester* at a naming ceremony on 23 September 1961. 46112 kept its plates for a few weeks after the naming of D100, then, unusually, not only had the nameplates removed but also the backing plate.

46112 Sherwood Forester (rebuilt September 1943) sits at Nottingham Victoria station during the period that the engine was allocated to Annesley shed, Nottingham. The engine lost its name (and the two plaques under the name) when a Peak diesel was given the same name. Unusually, when the nameplates were removed the backing plate was also removed, as can be clearly seen in this photo. The engine arrived at Annesley shed 1962, being amongst the first arrivals, and remained at the depot until withdrawal in May 1964, following major collision damage to the front of the locomotive. It was towed to Cashmores yard in Tipton for scrapping in September 1964. COLOUR-RAIL

The majority of the rebuilt Class 7s lost their nameplates in 1962 and by early 1963 most had been removed, although the removal was very haphazard. For example, 46157 *The Royal Artilleryman* still had plates in September 1963 and a small number carried plates until the autumn of 1964. Strangely, the rebuilt Jubilees and Patriots seemed to keep their plates longer than most of the Scots. Examples of plates still being carried include:

• 45736 *Phoenix*, 45512 *Bunsen*, 45522 *Prestatyn* and 46166 *London Rifle Brigade* still had plates after the application of the yellow stripe in August 1964
• 45527 *Southport* had plates in August 1964, but the crest above the name was missing and the yellow stripe had been applied
• 46155 *The Lancer* was seen with plates at Buxton shed in July 1964.

A number of steam locomotives had their plates removed when a new diesel locomotive was given the same name. Sometimes when a new diesel was named at a ceremony, the rebuilt Scots' removed regimental nameplate would be presented to the Regiment. For example, at the naming ceremony for Deltic D9000 *Royal Scots Grey* held on 18 June 1962, the Regiment was presented with the plates from 46101 *Royal Scots Grey*.

Similarly, at the naming ceremony for Deltic D9021 *Argyll and Sutherland Highlander* on 29 September 1963 at Stirling station, a nameplate from 46107 Argyll and Sutherland Highlander was presented for the regimental museum.

Other examples of rebuilt Class 7 names being reused on Deltics from rebuilt Scots include:

• D9008 was named *The Green Howards* on 30 September 1963 at Darlington
• D9010 *The King's Own Scottish Borderer* was named at Dumfries on 8 May 1965
• D9013 *Black Watch* was named at Dundee on 16 October 1963
• D9014 *The Duke of Wellington's Regiment* was named at Darlington on 20 October 1963.

The bulk of the regimental plates were gifted to the Regiments concerned (usually only one plate, but in some cases both plates) and many of these plates are on display at the appropriate regimental museums, whilst others are in the possession of the Regiment, but not necessarily on display. This is the case for 46112 *Sherwood Forester*, where the regimental museum has both nameplates and the plaques, but none is on public display. Following a request, the author was allowed to photograph both nameplates and also the plaque, which was carried below the nameplate detail-

46112 Sherwood Forester (rebuilt September 1943). The nameplates from the locomotive were presented to the regimental museum, where they remain today. The two crests carried above the name are also held by the Regiment and display distinct differences, the right-hand crest being inferior in quality and material from the one on the left and gives the impression that it has been cast not from a proper pattern, but using the other crest to cast from it. The locomotive had been officially named on 1 June 1933 at a ceremony at Derby station, when the crests were presented. The engine lost its nameplates on 23 September 1961, when the Class 45 diesel D100 (later 45060) was named Sherwood Forester at a naming ceremony at Derby station in the presence of 46112. The plates remained on 46112 for a few weeks, before they were removed and presented to the Regiment. When the nameplates were removed from the engine, unusually the backing plates were also removed.

ing the occasions that the loco pulled trains containing the Regiment.

By 1964–5, most of the locomotives had lost their nameplates, although a couple of engines were used on enthusiast 'specials' and replica wooden nameplates were made up at depots to smarten up the engines. Examples include 46115 *Scots Guardsman*, which spent some days prior to a 'special' in February 1965 being cleaned and having some replica wooden plates fitted at Crewe North shed (although at this time the loco was allocated to Carlisle Kingmoor). 46155 *The Lancer* had similar treatment when working a special train on 19 September 1964. However, 46155 had been seen in 1964 at Buxton with its original plates, but had also been seen without plates, so it may be that the original plates had been restored to the locomotive.

In 26 July 1964, 46115 *Scots Guardsman* was seen at Polmadie shed without nameplates, but with 'Everton FC' chalked on to the nameplate backing plate.

Front Smokebox Number Plates

The first rebuilds – the rebuilt Jubilees, Patriots and Scots – retained their LMS smokebox number plates, but with the formation of British Railways in January 1948, during the transitional phase before the uniform application of BR liveries, a number of variations were noted among the front smokebox plates as follows:

- Some engines had 'M' cast on a separate plate placed in front of the LMS number so that it would read 'M6138'. Examples observed include 6138 *The London Irish Rifleman* and 6170 *British Legion*.
- Some of the replacement BR number plates had a scroll and serif style of lettering, but were subsequently replaced by the standard BR style of lettering. Examples seen include 46149 *The Middlesex Regiment*.

Yellow Warning Stripe to Cab Side

With the activation of the 25Kv overhead wires south of Crewe on September 1st 1964 and the reduced clearances under some bridges a number of classes were banned south of Crewe and the crew's were warned by the painting of a yellow stripe painted across the number on the cab side.

The Derby Drawing Office issued a drawing (DD 5014) entitled 'Locomotives Prohibited from Working on Electrified Lines South of Crewe. Painting of Cab Sides', with the instructions: 'Locomotives to have a diagonal line 6in wide painted on the cab sides, the lowest point being adjacent to cab footstep. Lines to be painted yellow.' Most of the rebuilt Class 7s to receive the yellow stripe had it applied as per the directive, but in the case of 46160 *Queen's Westminster Rifleman* and 46166 *The London Rifle Brigade* the stripe

started and finished at the cab side lining, rather than coming from the corners of the cab. The yellow stripe was usually applied at sometime in August 1964, but no records exist of the actual dates of application, or to which locomotives, as it was the responsibility of the shed staff. The list below has been derived from studying numerous photographs:

- 45735 *Comet*
- 45736 *Phoenix*
- 45512 *Bunsen*
- 45522 *Prestatyn*
- 45526 *Morecambe and Heysham*
- 45527 *Southport*
- 45530 *Sir Frank Ree*
- 45531 *Sir Frederick Harrison*
- 46115 *Scots Guardsman*
- 46122 *Royal Ulster Rifleman*

46155 The Lancer (rebuilt August 1950) is seen displaying the yellow warning stripe, which all rebuilt Class 7s had to display from 1 September 1964 and which indicated that the engine was not to work under the overhead wires south of Crewe. In this case, the line went from one corner of the cab to the other, but as these warning stripes were applied at the depots sometimes the yellow line only went inside the panel formed by the cab side lining. R. ESSERY

45735 Comet (rebuilt May 1942) is seen at Annesley shed in 1964 with its nameplates removed and the yellow warning stripe applied to the cab side. The engine had arrived at Annesley in November 1963 from Willesden and was withdrawn from Annesley in October 1964. The engine retains the original flat top to the top feed and has been fitted with AWS (May 1959) and speedometer (December 1960). It possesses a riveted tender, which it had from rebuilding.

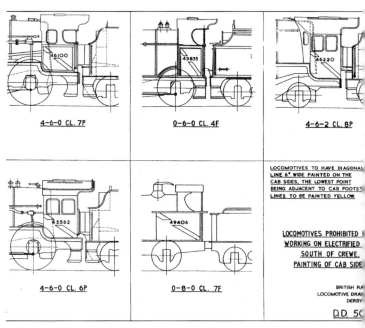

46166 London Rifle Brigade (rebuilt January 1945) is seen at Carlisle on 30 August 1964 and shows off its newly acquired yellow warning stripe across the cab side. The yellow line has not been applied as per the official instruction, as it should go from one corner of the cab to the other, but in this case it finishes where the lining is. The yellow stripe had to be in position by 1 September 1964, so most were applied at the depots in the last two weeks of August, with the earliest dating from 20 August 1964. 46166 remained at Carlisle for only a few more weeks, being withdrawn in September 1964. N. PREEDY

- 46125 *3rd Carabinier*
- 46128 *The Lovat Scouts*
- 46140 *The King's Royal Rifle Corps*
- 46148 *The Manchester Regiment*
- 46152 *The King's Dragoon Guardsman*
- 46155 *The Lancer*
- 46156 *The South Wales Borderer*
- 46160 *Queen's Victoria's Rifleman*
- 46162 *Queen's Westminster Rifleman*
- 46165 *The Ranger* (12th London Regt)
- 46166 *London Rifle Brigade*.

There is a degree of uncertainty over exactly which locomotives carried the stripe, as a number of them that would have been expected to receive the stripe were withdrawn at around the time it would have been applied. Certainly one loco, 46163 *Civil Service Rifleman*, was seen condemned at Annesley shed in September 1964 and part of the cab side had been cleaned (a diagonal panel just wide enough for the yellow stripe) to allow the shed painter to apply the yellow

This is the official BR drawing that was sent out to locomotive depots to show the shed staff how to apply the yellow warning stripe to certain classes of engine, which prohibited them from working south from Crewe from 1 September 1964. Despite being issued with this drawing, some applications of the yellow stripe did not conform to the drawing!

stripe, but the loco had been withdrawn before the stripe could be applied.

Revised Tender Emblem

The BR emblem on the tender was revised in 1957. Engines that went through Works after this date would receive the revised emblem if the tender was repainted, but at least three rebuilt Class 7s were withdrawn in the 1960s with the old emblem. Known examples are as follows:

- 45540 *Sir Robert Turnbull*, withdrawn in April 1963 from Carlisle Upperby
- 46113 *Cameronian*, withdrawn in December 1962 from Leeds Holbeck depot; the engine was observed in 1962 still with the early emblem
- 46159 *The Royal Air Force*, withdrawn in November 1962 and seen on the scrap line at Willesden in early 1963 with the early emblem.

46159 The Royal Air Force (rebuilt October 1945) sits in 1A Willesden shed on 25 November 1962, having just been withdrawn. The engine still has the first style of tender emblem, which for most engines was replaced from 1957 onwards with the revised version. This was one of three rebuilt Class 7s that carried the old emblem up to withdrawal. The nameplates have been removed, but all the motion is still in place as it was usual to take off the connecting rods when towing the engine away for scrapping. The engine moved to Crewe Works in November 1962 and was finally scrapped at Crewe in March 1963. RAILONLINE

Overhead Warning Indicators

From April 1960, following Continental practice, BR started to fit white enamel plates with the symbolic warning sign of forked lightning (in red) to strategic positions on the boiler cladding and tender to warn of possible contact with overhead wires. As the use of overhead wires for electrification extended, there were a number of accidents where steam loco crews came into contact with the overhead power lines when either trimming coal, putting water in the tender or fitting lamps. The position of these overhead warning flashes varied from engine to engine, but the usual placing was as follows:

- on the smoke deflectors near the bottom handrail
- rear of the tender
- on the firebox sides
- on the boiler sides

46169 The Boy Scout (rebuilt May 1945) is seen at Wolverhampton High Level station with an express sometime in 1959–60. It has a head code indicating that the train is from Edinburgh to Birmingham New Street. The engine shows off its Crewe North shed plate, where it was allocated between July 1959 and April 1962, before it moved to Willesden. The expresses between Wolverhampton and Euston were once the preserve of Bushbury, but in the early 1960s responsibility for providing the motive power passed to Crewe and Camden (and to Willesden upon the closure of Camden). The engine has AWS (fitted July 1959) and a speedometer drive (October 1960). The engine was named by Lord Baden-Powell on 9 December 1930 and a small badge can just be seen below the nameplate. The engine finished its days at Annesley, where it arrived on February 1963 but only lasted a few months, before being withdrawn in July 1963. It was scrapped at Crewe Works in August 1963.

46130 The West Yorkshire Regiment (rebuilt December 1949) is seen at Stockport Edgeley station ready for departure and displays the regimental crest above the nameplate. The rebuilt Class 7s were the principal express locomotives from Manchester London Road to Euston until the advent of the English Electric Type 4s (Class 40). The engine was allocated to Edge Hill between October 1957 and February 1959 and was then reallocated to Leeds Holbeck in November 1959. A short move to Low Moor followed in September 1961, before a return to Holbeck in June 1962 and withdrawal in December 1962. M. BENTLEY

46148 The Manchester Regiment (rebuilt July 1954) is seen at Birmingham New Street in ex-Works condition. The engine was allocated to 5A Crewe North in April 1962, before it moved to Llandudno Junction in September 1962, then on to Holyhead in December 1963. The engine is fitted with AWS (May 1959) and a speedometer (October 1960). COLOUR-RAIL

46148 The Manchester Regiment on shed at Willesden in August 1963 along with a Jubilee and another re built class 7 in the distance. At the time of the photograph the engine was allocated to 6G Llandudno Junction before moving to Holyhead in December 1963. The engine was withdrawn in November 1964. COLOUR-RAIL

CHAPTER FOUR

TESTING AND EXPERIMENTATION

As the group of engines was essential to the Operating Department any problems or issues required to be rectified as much as possible. Also the post-war need to reduce maintenance costs and reduce the amount of time for preparation and disposal was always driving the need to modify the engines.

Comparison Tests Crewe March 1945

In March 1945, rebuilt Jubilee 5735 *Comet* was used on Dynometer Car trials between Crewe and Carlisle, where it was compared to rebuilt Scot 6131 *The Royal Warwickshire Regiment* and original parallel boilered-Scot 6162 *Queen's Westminster Rifleman*. The objective was to determine coal and water consumption and steaming using two different sizes of blast pipe caps. The trials were run on three separate occasions, run-

ning on a regular Crewe to Carlisle service and return. The results showed that the rebuilt Jubilee with a 3¾in cap steamed freely, but it was noted that the riding was only considered to be fair, as lateral oscillation or nosing occurred at 65mph (105km/h).

Rough Riding

Almost from day one, the engines were reported for rough riding and extra maintenance was put into place to try to eliminate this problem. In 1947, a full-scale investigation was undertaken by John Powell, a loco inspector who spent two months at Longsight, Camden and Crewe North, checking lateral clearances on coupled axle boxes, axle-box top clearances, the condition of bogie slides and the condition of springs, as well as riding on the engines. His road tests revealed

46120 Royal Inniskilling Fusilier *(rebuilt November 1944) is seen on the Midland Main Line at Great Rocks in the Peak District near Buxton. It displays a hybrid livery, with the basic livery being the LMS 1946 livery, but with 'British Railways' on the tender and the BR number on the cab side. The smokebox number has LMS-style serif lettering, which was subsequently changed. The engine displays a 17A Derby shed plate, to which the engine was allocated from Longsight on four separate occasions between 1948 and 1949. The reason for the loan to Derby was for the locomotive to be tested for rough riding by the Locomotive Department. The engine is coupled to the LMS Dynomometer car and cables can be seen on the front of the engine and along the tender.* MIDLAND RAILWAY TRUST

TESTING AND EXPERIMENTATION ■

that some engines were worse than others and if the track was not in the best condition, the engines were more than just rough, usually resulting in the driver having to shut off steam and apply the brake.

Part of the problem was down to the very stiff bogie side-control springs, which would transfer any nosing from the front of the engine to its rear. Excessive wear on the faces of the rear axle boxes also made the situation worse and a temporary measure by Crewe Works to fit a gunmetal liner to the outer face of the axle-box face was a stopgap, but if this liner then dropped off in service (which they did regularly), it would not go unnoticed on the footplate. Powell's recommendations were to:

- use stiffer coupled springs
- stop using gunmetal liners on the axle-box face and use a more substantial method for building it up
- soften the bogie side control springs and introduce some form of friction-damping on the bogie slides; an experiment was issued to investigate the use of a hydraulic damper made by Monroe (well known for making shock absorbers for cars and motorcycles), but nothing else is known of this work.

The initial response to these recommendations was not good, but eventually a very senior person in the Motive Power Department spent three days on the footplate, returning with the opinion that the recommendations had some merit.

The result was that 46120 *Royal Inniskilling Fusilier* was allocated to Derby on four separate occasions between 1948 and 1949 (always 'on loan' and returning to 8A Longsight each time). The locomotive was tested between Derby and Buxton by the Locomotive Department, who wanted to measure forces on the bogie and axle boxes. The engine was coupled to the LMS Dynomoter car and measuring devices were connected to the front of the engine. Cables were run along the engine and along the tender to the instruments in the Dynomoter car. Changes were made and then retested. The result of these tests was that the best all-round results were obtained from bogie side springs giving only 1.5 tons of initial control (compared to 4–5 tons of the original springs) and all of the engines were subsequently modified.

The recommendation to fit stiffer coupled springs was also implemented, which helped to steady the riding, but at the same time made the engines very 'hard', with the result that some drivers swore you could feel every stone in the ballast.

Despite all this work, some engines seemed to refuse to improve – 46120 *Royal Inniskilling Fusilier* and 46131 *The Royal Warwickshire Regiment* were particularly bad and it was only after the driving wheels were analysed on the balancing machine at Crewe Works that the problem was identified. The original Fowler wheels had been 'tinkered' with by the addition of small auxiliary weights, but had never been spun up on the balancing machine to check. In addition, the coupling rods, which had also changed, had exaggerated the problem. Subsequently, all the locomotives had their wheel sets rebalanced.

Other experiments were listed as follows:

- M/C/L/1451 in 1958 – 'Bogie Bolster slides fitted with LA grade Ferobestos pads in an effort to improve riding'; fitted to 46120
- M/C/L/1479 in 1958 – 'Woodhead Monroe Shock Absorbers fitted to bogie side check gear with a view to improve riding'; fitted to 46120, but it would not take a genius to work out that a high-precision device such as a shock absorber would not take kindly to the sort of environment under the front of a steam locomotive (shock-absorber efficiency also falls off with increases in temperature)
- M/C/L/1451 in May 1959 – 'Bogie Bolster slides fitted with LA grade Ferobestos pads in an effort to improve riding'; again fitted to 46120 Royal Inniskilling Fusilier.

One of the worst engines identified for riding, as noted above, was 46120 *Royal Inniskilling Fusilier*, which was used for testing a number of improvements. It was loaned to Derby shed on the following dates: 12 February 1949 to 7 May 1949; 21 May 1949 until 11 June 1949; and 18 June 1949 until 3 December 1949. On each occasion, it went back to Longsight for a week or two before returning to Derby, presumably to test out changes on the West Coast Main Line. It was a 'guinea pig' again in 1955, arriving at Derby on 8 October 1955 and not returning to Longsight until 18 February 1956.

Smoke Deflectors

In 1947, a number of complaints were received regarding drifting smoke along the boiler sides obscuring the view ahead. At Bushbury depot, the senior drivers insisted that the rebuilt Patriots be moved away as a safety issue, even though the crews loved the engines. If the engines were running on the first valve of the regulator with a short cut of 15 per cent at high speed, the driver's vision was impaired. So, in 1947, 46115 *Scots Guardsman* was fitted with a set of deflectors (and was the only member of the class to have deflectors and carry the 1947 livery) and Powell was asked to ride on the engine to test their effectiveness. With running at 70mph (113km/h) on the first valve of the regulator and with 15 per cent cut off, he reported that the problem still existed. However, the Motive Power Department disagreed and the rest of the class was fitted with this style of deflector. The rest of the rebuilt Class 7 was not fitted with deflectors until September 1949 onwards, with the process taking until the middle of 1952. 46106 received the BR-style deflectors in July 1954.

Self-Cleaning Smokeboxes

As part of the drive to improve availability and to reduce the time taken to service engines, self-cleaning smokeboxes were developed, which in theory would mean that the smokebox would not have to be opened up for ash to be shovelled out on a daily basis. The key factor was, of course, that the steaming should not be impaired and that the screens fitted should not allow the smokebox to fill up with ash over the longer time intervals between the smokebox being opened.

In 1945, the new-build Stanier Black Five 4-6-0s had self-cleaning equipment fitted, but most of the class remained in service without the equipment and engines fitted would have an 'SC' on a small plate below the shed plate on the smokebox front. The problem was to make the equipment effective without causing steaming issues, a difficult trick to pull off.

The LMS was, of course, keen to fit the large passenger engines with these self-cleaning smokeboxes, but they were not a universal success when fitted to 3- and 4-cylinder engines. The 2-cylinder classes had a nice sharp bark, but the softer exhaust of the 3- and 4-

46151 The Royal Horse Guardsman *(rebuilt April 1953) is seen at Birmingham New Street with an express and the SC plate indicating a self-cleaning smokebox can be clearly seen under the shed plate of 5A. Only a small number of rebuilt Scots were fitted with the self-cleaning apparatus, as it proved difficult to make it work and give reliable steaming. This engine moved away from Crewe North in July 1959, so the photo predates that move.* N. PREEDY

46121 H.L.I. (rebuilt August 1946) is seen shortly after rebuilding and displays a very prominent 'SC' on the smokebox front to signify that it has self-cleaning apparatus inside the smokebox. At this time, only three rebuilt Class 7s were known to be fitted with this apparatus and further experiments took place in the 1950s, with revisions to the apparatus to minimize any adverse effects on steaming. The engine carries the original short form of the name before being renamed the Highland Light Infantry on 22 January 1949 and receiving a much bigger nameplate.

cylinder engines presented a new set of problems. Only a small number of Duchess Pacifics were fitted (46253, 46254 and 46255). In 1946, an experimental order was issued for ten Jubilees to be fitted, but it is not clear whether this happened. However, the LMS did not give up and two rebuilt engines were fitted with self-cleaning equipment, 6128 *The Lovat Scouts* and 6161 *King's Own*, both fitted with when rebuilt in 1946 and 6121 *H.L.I.* was noted with a painted 'SC' on the smokebox in 1946.

In the late 1940s, 46161 *King's Own* was fitted with an experimental set and was tested over a number of years. The equipment was based on that fitted to the Stanier Black Fives and the intention was that the smokebox would only need to be opened at every X examination or wash-out, which would normally be every eight to ten days. Charles Taylor, who was in the CME Experimental Department, reported on some runs where time had been lost by the engine. When the smokebox was opened, it was found to be full of ash, which would explain the poor steaming. The Crewe North shed foreman had the solution – the equipment was removed and dumped at the back of the shed.

However, the Locomotive Department did not give up and the Derby Design Office came up with a revised scheme, with 46101 *Royal Scots Grey* selected as the 'guinea pig'. The engine was at Camden at the time and

the Experimental Department did not receive any adverse comment as to the steaming of the engine. But when in September 1950 the engine moved to Crewe North and the complaints started again, one driver putting on the repair card 'SC stands for Short of Steam', there was a suspicion that the equipment had in fact been removed at Camden and put back in when the engine moved to Crewe. Whatever the reason, this version of the self-cleaning equipment was not consistent, so was removed. This failure is also recorded as being the reason that a similar experimental job for the Jubilees was closed down.

But the Locomotive Department persevered and the LM Region issued BR 358/16 in November 1953, entitled 'Experimental fittings on steam locomotives – Modified self-cleaning smokeboxes'. The following locomotives were listed: 46101, 46110, 46140, 46141, 46151, 46158, 46161, 46162, 46165 and 46167. In January 1958, another experiment was listed, using the same engines as in 1953. In May 1959, another experiment was listed for a modified self-cleaning smokebox, again with the engines listed in 1953 and 1958. Given the intensive diagramming of the rebuilt Class 7s, one can see why the CME Department wanted to improve the time taken for servicing, but it is clear that finding a workable solution for the rebuilt Class 7s was proving difficult.

46165 The Ranger (12th London Regiment) (rebuilt June 1952). At the end of 1955, 46165 was allocated to the Rugby Test Centre between February 1955 and June 1956 (although the EHC indicates that it spent two weeks at Carlisle Upperby in January 1956). The engine had already been authorized to be fitted with a modified self-cleaning smokebox at the end of 1953 and the tests were designed to see the impact on steaming. The engine is seen here with all the equipment to gather the data on smokebox vacuum, steam-chest pressures and so on. The testing plant has a full set of rollers inside the building to allow complete testing in a controlled environment. COLOUR-RAIL

The usual indication of having a self-cleaning smokebox, an 'SC' plate below the shed plate, does not seem to have been fitted in all cases, although photographs exist showing some engines so fitted in the mid-1950s, including 46151 *The Royal Horse Guardsman*.

46165 *The Ranger* (12th London Regiment) (rebuilt June 1952) was tested at Rugby towards the end of 1955 with a self-cleaning smokebox, but again it had a detrimental effect on the steaming. Alterations were made to the blast pipes and chimney liners, which improved evaporation by about 5 per cent. John Powell reported that ten Royal Scots were fitted with self-cleaning apparatus, but how many actually ran with the equipment is another matter.

Roller-Bearing Inside Connecting Rods

Mr Peter Skellon of the Bahamas Locomotive Society was the source of information relating to eleven locomotives being fitted with roller bearings to the inside connecting-rod big ends under experimental fittings reference M/C/L/1407 as follows:

- 46101 *Royal Scot Grey*
- 46114 *Coldstream Guardsman*
- 46120 *Royal Inniskilling Fusilier*
- 46125 *3rd Carabinier*
- 46128 *The Lovat Scouts*

- 46129 *The Scottish Horse*
- 46134 *The Cheshire Regiment*
- 46136 *The Border Regiment*
- 46152 *The King's Dragoon Guardsman*
- 46163 *Civil Service Rifleman*
- 46169 *The Boy Scout*.

46142 *The York and Lancaster Regiment*
45154 *The Hussar*
46157 *The Royal Artilleryman*
46165 *The Ranger* (12th London Regt)
46166 *The London Rifle Brigade*.

The first locomotive so fitted was 46169 *The Boy Scout* in July 1959 (when it was at Crewe Works for a Heavy General repair) and the last was 46152 *The King's Dragoon Guardsman* in March 1962.

Modified Draughting to Double Blast Pipes

Mr Peter Skellon also provided official information that a small number of rebuilt Scots received modified draughting to the double blast pipe and chimney under the experimental fittings reference M/C/L/1479. The locomotives concerned were as follows:

46108 *Seaforth Highlander*
46123 *Royal Irish Fusilier*
46134 *The Cheshire Regiment*

The 1948 Locomotive Exchange Trials

With the formation of British Railways in 1948, the future motive-power strategy now came under one Chief Mechanical Engineer, rather than the independent engineers of the big four companies. As part of this process of change, it became necessary to determine whether it would be a case of carrying on as before with each of the regions being given relative independence, or whether the newly formed British Railways would design a new range of locomotives incorporating 'best practice' from each of the former big four companies. The end result was that BR decided to build new 'standard' classes as a stopgap before extensive electrification would take place. The Locomotive Exchange Trials took place between April and September 1948 with three categories – express passenger, mixed traffic and freight.

46154 The Hussar (rebuilt March 1943) is seen at Waterloo station during the 1948 Locomotive Exchange Trials. Because the Southern Region did not have water troughs and the water capacity of the tender of the rebuilt Scot was considered marginal, a tender from a WD 2-8-0 has been borrowed and given LMS lettering, despite the fact that the engine now bears its BR number along with the LMS 1946 livery. The engine did not receive its smoke deflectors until October 1951.

The rebuilt Scots were placed in the express passenger group along with the Princess Coronation Pacifics and they would be compared to the ex-GWR Kings, the ex-LNER A4s and the Southern Region Merchant Navies (the ex-SR West Countries were placed in the mixed-traffic category). Although technically a lower power classification than the GWR Kings, the grate area at 31.3sq ft (2.9sq m) was only marginally smaller than the Kings 34.3sq ft (3.2sq m) and the higher superheater value would claw back some of the disadvantage. Two rebuilt Scots were selected – 46154 *The Hussar* and 46162 *Queen's Westminster Rifleman*.

Because of the absence of water troughs on the Southern Region, a 5,000gal (22,730ltr) tender from a WD was fitted to 46154 *The Hussar* (as against the standard 4,000gal [18,184ltr] tender more usually carried).

The two rebuilt Scots ran on all the regions as follows:

- Western Region – Paddington to Plymouth and return
- Eastern region – King's Cross to Leeds and return
- Midland Region – Euston to Carlisle
- Southern Region – Waterloo to Exeter and return.

Steaming Tests 1955

46165 *The Ranger* (12th London Regiment) (rebuilt June 1952) was tested at Rugby towards the end of 1955. The self-cleaning smokebox was found to have a detrimental effect on the steaming. Alterations were also made to the blast pipes and chimney liners, which improved evaporation by about 5 per cent.

Testing 1956

46120 *Royal Inniskilling Fusilier* was tested once again over the Peak District in 1956 with the Dynamometer car; as before, it was allocated to Derby between October 1955 and February 1956.

46167 The Hertfordshire Regiment *(rebuilt December 1948) is seen at Willesden shed in 1963 when the engine was allocated to Willesden, having arrived in June 1963 before moving to Annesley in September 1963. It was withdrawn on April 1964 and scrapped at Crewe Works in May 1964. The engine has AWS (July 1959) and a speedometer drive (September 1960) and has Stanier bogie wheels, but still has the original Fowler wheels. The tender is the one fitted when the engine was rebuilt and this was retained until final disposal.*

46164 The Artists' Rifleman (rebuilt June 1951) is seen at Willesden in April 1955 after working a Rugby League special train to Wembley, where, after dropping off the train and passengers, the loco would be serviced at Willesden. The use of such headboards on special trains was not unusual in the 1950s and the engine would have been sent to Barrow shed the day before to be prepared for its trip. The engine has yet to be fitted with AWS and was allocated to Edge Hill at the time, having arrived at the depot in February 1942 and not leaving for Crewe North until September 1959.

46103 Royal Scots Fusilier (rebuilt June 1943). The first rebuilt Scot sits at Crewe in the early 1950s with an express. It has a 20A Leeds Holbeck shed plate (later recoded to 55A in January 1957), so is a long way from its home depot and would have been normally seen working north of Leeds to Carlisle and Glasgow.

CHAPTER FIVE

ALLOCATIONS

Because the rebuilding process was quite slow it would take a few years for the rebuilds to have a major impact on the London Midland and Scottish railway, but the impact on individual depots, particularly in the war years, was considerable. World War II placed huge burdens on the railways with not only the increase in freight traffic to service the needs of the military, but also the need to increase passenger capacity by adding additional coaches to trains. Passengers recount tales of trains being so long at Euston that the locomotive was no longer at the platform end, but halfway up Camden Bank!

The first rebuilt Jubilee, 5736 *Phoenix*, was allocated to Leeds Holbeck shed following the rebuilding process in May 1942, initially on loan and then permanently a few weeks later. It was followed by 5735 *Comet* in June, again initially on loan and then permanently. It is assumed that during the loan period the locomotives would be the responsibility of Crewe Works and the Chief Mechanical Engineers Department and they would be looking to see if there were any serious issues. Any concerns that they might have had were unfounded, as the Holbeck crews welcomed the locomotives with open arms, nicknaming the two engines 'the heavenly twins'.

The engines worked on the heavily graded route over the Settle and Carlisle line to Carlisle and over Beatock to Glasgow. They remained at Leeds Holbeck only for a short period before being reallocated to Camden at the end of July 1942, where they would be faced with the enormous trains prevalent during the war years from Euston. Leeds Holbeck shed was only without a rebuilt Class 7 for a few weeks, as 6170 *British Legion* was loaned in July 1943 and stayed for a

month before moving to Camden. The depot received the first rebuilt Scot in the shape of 6103 *Royal Scots Fusilier*, which arrived at the shed at the end of August 1943 and was to remain there until 1958.

The next shed to receive rebuilds was Camden, which was the principal passenger depot for London Euston.

The allocations of rebuilds was coloured by the frequent transfer of certain locomotives and the use of 'loans' to depots for short periods to cover the non-availability of other rebuilds (such as Works visits or repairs, or running on lesser trains to accumulate mileage for a Works visit). At the peak of steam operations, there was only just enough rebuilt Class 7s at many depots (indeed, some railwaymen were of the opinion that there were never enough), so the sudden non-availability would mean a request to Control to send a substitute for a few weeks. The rebuilt Class 7s were also on very intensive diagrams, which would mean high mileages being racked up. As an example, Kentish Town would send a rebuilt Scot to Manchester Central from St Pancras and the loco would then work an express to Buxton, before returning to Central and then back to St Pancras.

Some of the rebuilds spent a short amount of time at a number of depots, such as 46125 *3rd Carabinier*, which in its career as a rebuild was reallocated thirty times, appearing on Crewe North's books as many as thirteen times. Not only were the engines being worked on intensive diagrams, but the loads being hauled were also usually heavy. For example, it was not unusual to see a rebuilt Class 7 with fifteen or more coaches and if substituting for a Duchess the load could be up to nineteen coaches. However,

46122 Royal Ulster Rifleman (rebuilt September 1945) is seen at Camden in 1962. It is seen in its final condition and has been fitted with Stanier bogie wheels and AWS (October 1959) and a Smith-Stone speedometer drive (September 1960). It retains its original riveted tender. The overhead warning flashes can also be seen, which would have been fitted from 1960. There was a small crest below the nameplate fixed to the splasher. The engine was transferred to 16D Annesley in December 1962 and remained there until October 1962, when it was nominally allocated to 12B Carlisle Upperby, but was withdrawn in the same month and was stored at Carlisle Upperby until January 1965. It travelled to Drapers of Hull for scrapping in February 1965.

46149 The Middlesex Regiment (rebuilt April 1945) is seen at Birmingham New Street in the early 1960s. It was a 6J Holyhead engine until April 1960, when it went to 5A Crewe North, then to 9A Longsight in June 1961. It was withdrawn from that depot in September 1963 and was scrapped at Crewe Works in October 1963, after being stored in the Works for a few weeks. COLOUR-RAIL

46135 The East Lancashire Regiment (rebuilt January 1947) is seen having just been coupled up to the midday Scot at Crewe, with the fireman climbing back into the cab. The preparing crew had made sure that they had enough coal in the tender, as the coal is higher than the cab roof. It was standard practice at Crewe North shed to top up the tender when leaving the depot, as the fireman would have filled the firebox before departure and it would be a hard slog to Carlisle and Glasgow. The engine was a Crewe North engine between February 1957 to August 1962, with a couple of short moves to Camden and Longsight in-between. Crewe seems to have been a 'clearing house' for rebuilt Class 7s, with engines moving in and out of the depot regularly.

46157 The Royal Artilleryman (rebuilt January 1946) is seen at Camden, displaying its 21A Saltley shed plate, to where it had been allocated in June 1961. It was transferred away in June 1963 to Carlisle Upperby. The locos at Saltley could be seen on anything from an unfitted coal train up to an express passenger train. In the case of 46157, it is likely to have been a passenger train, as Camden only serviced locos that had worked into Euston; if it had arrived on a freight, it would have been serviced at Willesden.

other locomotives, such as 46104 *Scottish Borderer*, would spend its entire career as a rebuilt Scot at Polmadie depot, arriving in March 1946 until withdrawal at the end of December 1962. Most of the other rebuilt Scots at Polmadie, such as 46102 *Black Watch*, 46105 *Cameron Highlander* and 46107 *Argyle and Sutherland Highlander*, also spent their entire rebuilt career at the depot.

At depots such as Camden, if drivers complained about rough riding or poor steaming from an engine it would be sent up to Crewe North and a substitute engine would be delivered to Camden. Camden had no space to park up engines wanting repair or attention, so locomotives would be sent either to Willesden or back to Crewe, but a replacement would always be required. Other depots with rebuilt Class 7s also had 'loans'. For example, 46147 *The Northamptonshire Regiment* (6J Holyhead) was reallocated to 8A Edge Hill on 22 June 1957, but was loaned back to Holyhead the same day and was officially transferred back to Holy-

head on 21 September 1957, all without apparently leaving the depot. The loco was then loaned to 12A Carlisle Upperby in week ending 19 October 1957 and returned to Holyhead in week ending 23 November 1957. 45528 *REME* was a Crewe North engine when it was loaned to 6J Holyhead in week ending 13 April 1957, then returned to Crewe North in week ending 20 April 1957 for a stay of one week. The weekly movement sheets for the LM Region hardly had a week go by without a rebuilt Class 7 being reallocated or loaned to another depot.

Also, on the LM Region everything was based on mileage, so the premier trains would be given to the engines with the lowest mileage and the higher mileage examples would be moved on, or put on lesser workings until a Works visit could be arranged. Crewe Works was also adept at providing repairs in a relatively short amount of time, so a Heavy General, which meant that the boiler was taken off the frames, usually took about four weeks (and the boiler being

Introduction of Rebuilt Class 7s

Class	1935	1942	1943	1944	1945	1946	1947	1948	1949	1950	1951	1952	1953	1954	1955	Total
Jubilee		2														2
Patriot						2	6	9	1							18
Scot	1		9	9	11	9	4	5	6	6	2	2	4	2	1	71
Total	**1**	**2**	**9**	**9**	**11**	**11**	**10**	**14**	**7**	**6**	**2**	**2**	**4**	**2**	**1**	**91**

46148 The Manchester Regiment (rebuilt July 1954) is seen at Carstairs on 3 June 1963 with an express. The engine is a long way from its home of Llandudno Junction at the time, but it shows how during the transition to diesels engines would be borrowed by depots far from their base for passenger workings and it is likely that the engine came on to the train at Crewe. The engine moved around between Holyhead and Llandudno Junction, before being withdrawn in November 1964 from Holyhead. COLOUR-RAIL

replaced from the pool of spare boilers), minimizing the time away from the Operating Department. The Operating Department tried, where possible, to get heavy repairs done during the winter period, giving the largest possible pool of engines for the summer timetable, when they would run many extra trains. Crewe North seemed to act as a 'clearing house' for the rebuilt Class 7s and it would also be the depot that would release rebuilt Class 7s from Works visits and would be responsible for running-in the engine before it was sent off to a depot.

As the rebuilds were released from Crewe, they could be allocated to any of the principal depots of the London Midland Region (with the exception of the Midland Main Line until the 1960s), such as Camden, Crewe North, Longsight (Manchester), Edge Hill Liverpool, Holyhead, Holbeck (Leeds), Carlisle Upperby and Polmadie. It was only later in their careers that the rebuilds would appear at places like Nottingham, Derby and so on. The rebuilt Class 7s were able to take over the heavier trains worked by the Jubilees and the original Royal Scots and also provided back-up to the Princess Royals and the Princess Coronations. For details of when they were introduced, see the table.

The pace at which the rebuilding process happened was dictated by the life expectancy of the original boilers in the classes being rebuilt, so only locomotives that required a new boiler would be considered and as the pool of boilers for the parallel Scots and Patriots increased, the rate of conversion slowed and many original Scots received Heavy General overhauls instead of rebuilding in the early 1950s.

To summarize the allocation of the rebuilt Class 7s, the table gives two snapshots of the allocations in May 1957 and July 1962 and clearly shows the effect by 1962 of the introduction of the English Electric and Sulzer Type 4, and of course the rundown of Camden. 1962 also marked the withdrawal of the first of the rebuilt Patriots.

Looking at another snapshot in time, the following list shows the allocations as of January 1954 on the London Midland Region. At this point, the rebuilding process had almost been completed, with only three engines remaining to be rebuilt. This represents the high point, as the diesel invasion had not started and the rebuilds were concentrated at the 'premier'

Allocations of Rebuilt Class 7s in May 1957 and July 1962

	May 1957				July 1962			
	Rebuilt Jubilee	Rebuilt Patriots	Rebuilt Scots	Total	Rebuilt Jubilee	Rebuilt Patriots	Rebuilt Scots	Total
Camden	1	5	9	15				
Crewe North	1	2	24	27	1		11	12
Edge Hill		6	8	14		2	4	6
Longsight		3	10	13			2	2
Holyhead			2	2		1	5	6
Carlisle Upperby		3	7	10		5	16	21
Leeds								
Holbeck			8	8			5	5
Polmadie			5	5			5	5
Willesden					1	4	10	15
Nottingham							2	2
Llandudno Junction						2	3	5
Newton Heath						1	4	5
Trafford Park							3	3
Millhouses						1	2	3
Springs Branch						1	1	2

depots. The list of course does not include any engines at Glasgow Polmadie as this was part of the Scottish Region (as was Carlisle Kingmoor).

- Camden   Rebuilt Patriots (five): 45514, 45522, 45523, 45532 and 45545; rebuilt Jubilees (two): 45735 and 45736; rebuilt Scots (twelve): 46100, 46116, 46126, 46139, 46141, 46142, 46144, 46147, 46154, 46162, 46168 and 46170.
- Crewe North   Rebuilt Patriots (three): 45528, 45529 and 45535; rebuilt Scots (fourteen): 46101, 46106, 46118, 46119, 46128, 46134, 46138, 46140, 46151, 46155, 46159, 46163, 46166 and 46167. Crewe also had the three remaining non-rebuilt Scots: 46137, 46148 and 46156.
- Edge Hill   Rebuilt Patriots (five): 45521, 45525, 45527, 45531and 45534; rebuilt Scots (eight): 46123, 46124, 46135, 46149, 46152, 46153, 46158 and 46164.
- Longsight   Rebuilt Patriots (three): 45530, 45536 and 45540; rebuilt Scots (twelve): 46111, 46114, 46115, 46120, 46122, 46125, 46130, 46131, 46143, 46169, 46161 and 46169.
- Holyhead   Rebuilt Scots: (six) 46110, 46127, 46129, 46132, 46150 and 46157. The allocation of rebuilt Scots was a sizable proportion of the depot, as there were only nineteen engines in total allocated, with six of them being rebuilds.
- Carlisle Upperby   Rebuilt Patriots (two): 45512 and 45526; rebuilt Scots (three) 46136, 46146 and 46165.
- Leeds Holbeck   Rebuilt Scots (eight): 46103, 46108,

46109, 46112, 46113, 46117, 46133 and 46145.

Rebuilt Class 7s at the London Midland Region Depots

Now we will take a look at the individual depots, the rebuilt Class 7s that were allocated to them, plus the services that they were used on, starting with Leeds Holbeck as this was the first depot to receive the two rebuilt Jubilees.

Leeds Holbeck

Holbeck was coded 20A and was London Midland Region until it was transferred to North Eastern Region in January 1957 as 55A. It closed on 30 September 1967. It became famous for its 1965–7 stud of Jubilees, which became something of a celebrity.

The first report on the rebuilt engines was their use on Leeds to Bristol trains and the *Railway Observer* commented that: 'they were far superior and did not have to be thrashed'. On 16 November 1942, 5736 *Phoenix* was noted working the 5.23 from Birmingham New Street to Blackwell (on the route to Bristol) loaded to twelve vehicles and the driver stated 'she is a lovely engine, one that would pull back lost time'.

One of the initial workings for the two locomotives was the 6.43 from Carlisle to Leeds; 185min was allowed for the 110 miles (177km). The train was loaded

46103 Royal Scots Fusilier *(rebuilt June 1943) is seen in store after being withdrawn at 55C Farnley Junction shed in 1963. The first of the rebuilt Royal Scots spent many years allocated to Leeds Holbeck before a three-year stint at 14B Kentish Town (between December 1958 and July 1961), then short stays at 21A Saltley and Carlisle Upperby, before coming back to Holbeck in September 1962. Withdrawal came in December 1962 and the engine was placed in store at 55C Farnley Junction (Leeds) until August 1963, before being scrapped at Crewe in September 1963. A number of rebuilt Scots came back to Holbeck in 1962 to cover for the high failure rate of the Sulzer Type 4s (Class 45/46) and three examples were then stored at Farnley after withdrawal (46103, 46130 and 46145). The nameplate and crest have been removed, but the smokebox number plate is still in place and the Holbeck shed code has been painted at the bottom of the smokebox door. The bogie wheels are of the Stanier type, while the driving wheels are of the Fowler type. G. SHARPE*

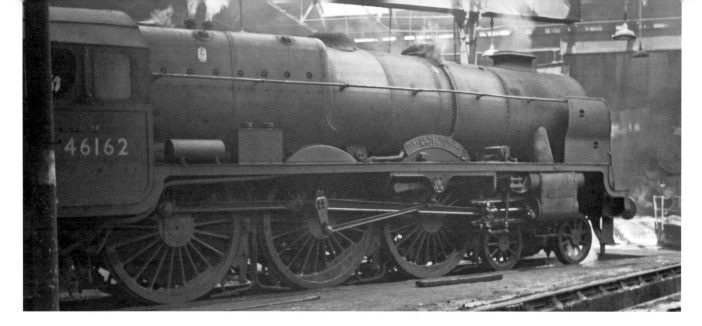

46162 Queen's Westminster Rifleman (rebuilt January 1948) sits in the shed at Leeds Holbeck on 18 April 1964 and unusually still retains its nameplates at this late date. The engine is a visitor from Carlisle Upperby, having been allocated to the depot in July 1962. It arrived with twelve other rebuilt Class 7s from the mass clear-out at Saltley, which itself had received the engines en masse from various Midland depots, such as Nottingham and Kentish Town, in June and July 1961. The engine was one of the last operational rebuilt Class 7s, being withdrawn in June 1964 and scrapped at Connels, Calder, in September 1964. The rebuilds at both Carlisle Kingmoor and Upperby were regular visitors to Leeds in 1964–5, when they could be seen working on stopping passenger services and freights between the two cities. The AWS tank can be seen, as well as the original Fowler-style driving wheels. KEN TYLER

Leeds Holbeck Shed (55A) on 18 April 1964, showing Jubilee 45608 Gibraltar, a long-term Holbeck resident still with its nameplates, rebuilt Scot 46162 Queen's Westminster Rifleman (Carlisle Kingmoor) and an 8F 48399 sitting around the turntable. Leeds Holbeck was long associated with the rebuilt Class 7s, with the two rebuilt Jubilees arriving at the shed following conversion in 1942. These were followed by the first rebuilt Scot, 6103 Royal Scots Greys, when it was converted in 1943 and a number of other rebuilt Scots followed. In 1962, all of Holbeck's rebuilds were transferred away to Low Moor and Mirfield, but a number returned in June 1962. In addition, 46161 King's Own from Crewe North arrived, but were all withdrawn at the end of 1962. KEN TYLER

46112 Sherwood Forester (rebuilt September 1943) is seen leaving Carlisle with the Thames–Clyde Express, which was a named train from St Pancras in London to Glasgow and return, via Leeds and the Midland Main Line. The engine is working hard, with the sanders on in the greasy conditions. The safety valves are just sizzling, ready for the slog ahead, either north over Beattock, or south over the Settle–Carlisle route. Sherwood Forester had arrived at Leeds Holbeck in February 1953 and remained until a transfer to Nottingham in December 1959, before a move to Trafford Park in December 1960 and then its final move to Annesley in February 1962. The engine was withdrawn in May 1964, following a collision with a Black Five in the dark. The engine was one of the few rebuilt Class 7s that was never fitted with AWS. G. SHARPE

at 550tons (558,000kg), consisting of eight coaches, one luggage van and eight to ten loaded milk tankers. It had five stops and as the route was very steeply graded this was a good performance. The two rebuilt Jubilees were referred to as 'the heavenly twins' and the crews at Holbeck would not hear a bad word against them. They were also of the opinion that the two rebuilt Jubilees were less lively at speed than the rebuilt Scots, although there was no concrete evidence to confirm this. The stay of the two rebuilt Jubilees was for a year, as the engines were transferred away to Camden at the end of July 1943. 6170 *British Legion* was in residence for a few weeks before the first rebuilt Scot, 6103 *Royal Scots Fusilier*, arrived, followed quickly by 6108 *Seaforth Highlander* in September 1943 and further allocations of newly rebuilt Royal Scots. The depot had eight rebuilt Class 7 locomotives to cover five diagrams.

The rebuilt Class 7s at the shed settled down and worked on the Leeds–Glasgow services and were rarely seen south of Leeds, these remaining the preserve of the depot's Jubilees, which worked services to Bristol (via Birmingham) and St Pancras in London. The Holbeck rebuilt Class 7s would work to Glasgow and back and only if there was a problem at the Glasgow end would any of Polmadie's rebuilds work into Leeds. The passenger trains on this route also included some heavy night-time sleeper trains.

On 10 July 1943, the pioneer loco 46170 *British*

Legion was loaned to the shed until 21 August 1943 when it went to Camden. It was believed that the loco was sent to Leeds for comparisons with the two rebuilt Jubilees and worked the same services on the heavy trains to Glasgow via Carlisle but in fact the engine missed being at Holbeck with the two rebuilt Jubilees by a few days, although it would have been possible to compare records of performance of the respective engines.

One of the peculiarities of Holbeck was the practice of cleaning a rebuilt Scot before it was sent to Crewe Works for overhaul and giving the engine a boiler wash-out when the locomotive returned from Works before being put into service.

The reign of the Holbeck rebuilt Class 7s on the principal workings to Glasgow came to an end not because of the allocation of any new diesels, but with the allocation of a number of ex-LNER A3 Pacifics in February 1960, when 60038 *Firdausi* and 60077 *The White Knight* arrived at Holbeck, followed by 60080 *Dick Turpin*, 60082 *Neil Gow*, 60088 *Book Law* and 60092 *Fairway* on 14 May 1960. On 26 November 1960, the Holbeck stud of A3s was supplemented by 60069 *Sir Visto*, 60070 *Gladiateur* and 60072 *Sunstar*. As the depot had been part of the North Eastern Region since March 1957, the allocation of ex-LNER Pacifics was an easy matter to arrange. The allocation of the Pacifics came about as there had been continual complaints from the travelling public about the late running of many trains and the Pacifics could be considered a better stopgap until the diesels arrived. The A3s only remained until July 1961, when the depot began to receive an allocation of the Sulzer Type 4 diesels (Classes 45 and 46).

With the allocation of a large number of Class 45 Peaks to the depot in 1960, in 1961 all the rebuilt Class 7s were transferred away to Low Moor (Bradford), but the speed of introduction and the inexperience of both crews and fitting staff meant poor availability, with the result that steam had to be pressed back into service and the shed had to 'borrow' locos from wherever it could find them. The close proximity of Low Moor shed and subsequently Mirfield meant that it was easy for Holbeck to 'borrow' some of its old locos back and as an example 46113 *Cameronian* was seen on 10 May 1962 on a Leeds–Morecambe express.

In March 1962, with the shed yard packed with Peaks waiting for repair or examination, a Crewe North rebuilt Scot, 46127 *Old Contemptibles*, was being prepared to haul the southbound Thames–Clyde express.

On 12 May 1962, 46130 *The West Yorkshire Regiment* (Low Moor) worked the 8.50 Leeds–St Pancras express, deputizing for a diesel.

In June 1962, all of the five locos that had been exiled to Low Moor returned to Holbeck to cover for diesel failures and for one last hurrah. However, their stay was short-lived and once the end of the year was reached their time was up and they went into store and withdrawal. Holbeck also had to lend engines, so 46136 *The Border*

Regiment was loaned to the depot from 5A Crewe North between August 1962 and September 1962. Following withdrawal, engines were placed in store and on 9 March 1963, 46113 *Cameronian*, 46117 *Welsh Guardsman* and 46161 *King's Own* were observed stored at Leeds Neville Hill and in January 1963, 46103 *Royal Scots Fusilier*, 46130 *The West Yorkshire Regiment* and 46145 *The Duke of Wellington's Regiment* (West Riding) were seen stored at Leeds Farnley Junction shed.

On 23 September 1963, 46145 *The Duke of Wellington's Regiment* (West Riding) was steamed for the last time for the melancholy task of pulling classmates 46103 *Royal Scots Fusilier* and 46130 *The West Yorkshire Regiment* to Crewe Works, where all three locomotives were scrapped in December 1963.

Even as late as 1966, just thirteen of the twenty-five Peaks allocated to the depot were available for service, resulting in the depot's small allocation of Jubilees being pressed into service (as the rebuilt Class 7s were long withdrawn).

Crewe North

Crewe North was coded 5A until its closure on 24 May 1965. It was one of the principal express passenger depots on the London Midland Region (Crewe South 5B was the freight depot for the town), with an extensive allocation of Princess Royals and Coronation Pacifics (around twenty Pacifics), as well as

46141 The North Staffordshire Regiment *(rebuilt October 1950) is seen at Crewe station with the pre-1957 British Railways emblem on the tender. The locomotive has acquired a full set of Stanier-style wheels with bolted-on balance weights and a set of Stanier bogie wheels. The top feed is the later style and the locomotive acquired its smoke deflectors in November 1950, shortly after its rebuild. The engine was withdrawn from Carlisle Upperby on April 1964 and scrapped at Crewe Works a few months later.* G. SHARPE

46125 3rd Carabinier (rebuilt August 1943). An official view taken sometime around the mid-1950s, the emblem on the tender is the pre-1957 version and the engine is yet to be fitted with either its speedometer drive or AWS and displays its Crewe North (5A) shed plate. The top feed is the later Ivatt type and the engine has yet to be fitted with the additional handrail at the bottom of the smoke deflectors and has a mixture of Stanier and Fowler bogie wheels. In 1963, the engine was seen with its nameplates, but without the crest on the side shown in the photograph. The locomotive was allocated to Crewe North on four separate occasions in the 1950s, ending its days at Annesley, where it was allocated from September 1963 until withdrawal in October 1964, before scrapping at Cashmores Yard at Great Bridge, Tipton.

rebuilt Class 7s, Jubilees and Patriots. In its heyday, the depot employed several hundred drivers.

The station was at a junction for lines to North Wales via Chester, north to Preston and Scotland, to Shrewsbury, Liverpool, Manchester and Stoke, a total of seven separate traffic flows – busy indeed. Many trains in the steam era changed engines at Crewe on all these routes. For example, the author well remembers standing at the station in 1962 and being amazed at the number of locos being changed, with a constant stream of locos coming to and from the shed. Crewe North was also responsible for many heavy overnight sleeper and mail trains between Crewe and Perth, plus the trains had coaches added or taken away as part of the junction's function.

The rebuilt Class 7s were used on trains to Birming-

ham, Euston, Workington, Windermere, Holyhead, Perth and Glasgow. Its engines were regularly seen on the North and West route (Shrewsbury, Hereford and Pontypool Rd), with Manchester to Kingswear and Liverpool to Kingswear services in the summer as well as Liverpool to Plymouth services. Another example of one of Crewe North's less likely workings was the 6.35 Crewe–Camden Class C milk and perishables train, which would be regularly hauled by a Class 7 rebuild.

The intensive nature of the diagrams for the rebuilt classes meant that a loco coming into the depot for disposal had so little time before going off shed again that the incoming crew would be carrying out the disposal (emptying the ash pan, cleaning the fire and so on), whilst the crew working the engine off the depot would be doing the preparation work (oiling round,

45526 Morecambe and Heysham *(rebuilt February 1947) is seen at Crewe after August 1964, displaying the yellow stripe through the cab side to indicate that the locomotive was banned from working south of Crewe due to the restricted height of the overhead wires. In most cases, the stripe went from one corner of the cab to the other, but in other instances (such as 46160* Queen Victoria's Rifleman*), the stripe only went from the edge of the cab side lining. The engine has also had the top lamp bracket moved to the 3 o'clock position and the lamp bracket on the platform above the buffer beam has been also relocated. The engine still retains its nameplate at a time when most rebuilt Class 7s were losing their nameplates. The rebuilt Patriots kept their nameplates longer than their rebuilt Scot compatriots. The engine is fitted with a welded tender, which most of the rebuilt Patriots carried.*

45534 E. Tootal Broadhurst *(rebuilt December 1948) is seen stored 'serviceable' at Crewe North shed on 8 March 1964. The term 'serviceable' meant that the engine could be steamed at any time as it was in a serviceable condition, which was usually indicated by sacking over the chimney to prevent rainwater going down the blast pipe. The engine has AWS (April 1959) and a speedometer drive (September 1960). It still retains its nameplates and also has a badge holder on the cab side for the driver's name. The locomotive had arrived at Crewe in June 1963 and was withdrawn May 1964, having been returned to traffic a few days after the photograph was taken (on 12 March 1964). After withdrawal, it made the short journey to Crewe Works, where it was scrapped in June 1964.* KEN TYLER

coaling up and filling the tender with water) at the same time so that a loco could arrive at the depot and be back off shed to the station in 2½ hours.

As many of the workings from Crewe were long distance, it was standard practice for the fireman to fill up the firebox (which would take about 1ton [1,016kg] of coal) and make a visit to the coaling plant to top up the tender when leaving the shed to pick up

a train. Photos of engine changes at Crewe reveal these very full tenders.

Crewe North was the largest passenger depot on the LMS and a summary of engines allocated to Crewe North can be found in the table below.

Crewe North seemed to act as a 'clearing house' for rebuilt Class 7s, with locos moving in and out of the depot very often being loaned to other depots for

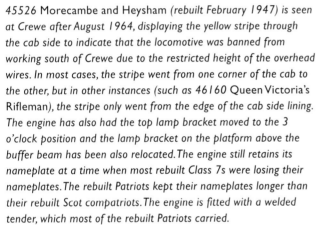

Summary of Engines Allocated to Crewe North

Year	Total Engines	Rebuilt Class 7
1950	95	11 (includes some non-rebuilt engines)
1956	110	22
1960	108 (plus 17 Diesels)	22

46170 British Legion (rebuilt October 1935) is seen at Crewe North depot after an overhaul in either 1960 or 1961. It is fitted with AWS and shows not only the air tank, but also the battery box on the platform, which was normally not visible on the rebuilt Class 7 locomotives. The distinctive nameplate is visible and has a red painted background. The top feed on the boiler top is located closer to the dome and the additional length of the No.2 boiler compared with the 2A boilers is noticeable by the overhang of the smokebox. RAILONLINE

45512 Bunsen (rebuilt July 1948) is seen at Crewe North depot sometime in the early 1960s. It had been a long-time resident at Carlisle Upperby, arriving there in 1958 and remained there until a transfer to Carlisle Kingmoor in November 1964. It was withdrawn in March 1965, being one of a very small group that survived into 1965. RAILONLINE

short periods before returning to Crewe North. There was a constant stream of engines being moved to and from Camden and Holyhead, sometimes only for a few weeks, but it is indicative of the intensive diagramming for the rebuilt Class 7s with no 'spare' engines in the system. Only the introduction of the English Electric Type 4s brought an end to the constant reallocation of engines, although Holyhead was still the recipient of Crewe's rebuilds until 1963.

Some engines were allocated to Crewe North a number of times, with 45528 *REME* being allocated on eight separate occasions, 46119 *Lancashire Fusilier* on nine occasions and 46125 *3rd Carabinier* thirteen times between 1943 and 1963, including a spell of eight years at the depot between 1954 and 1962. The record holder was 46166 *London Rifle Brigade*, which was allocated to the depot fifteen times between January 1945 and July 1961. The depot was also responsible for receiving locos from Crewe Works and running them in before they would be transferred to whatever depot they were allocated to.

By November 1961, twenty-seven EE Type 4s had been allocated to the depot and had taken over most of the depot's Class 7 and Class 8 workings, particularly to London. In the winter of 1963, the majority of rebuilt 7s at Crewe were put into store, with the duties being covered by the Britannias allocated to the depot. By September 1964, all the Class 8 Duchess Pacifics had been withdrawn and replaced by Britannias, further reducing the regular work of the rebuilt Class 7s. By September 1964, the last rebuilt Class 7 had left the depot.

The last two rebuilt Class 7s allocated to Crewe North were 45534 *E. Tootal Broadhurst*, withdrawn in May 1964, and 46155 *The Lancer*, which was transferred away in September 1964 to Carlisle Kingmoor. However, some of Carlisle's rebuilt Class 7s continued to visit Crewe North into 1965.

The depot also prepared 46115 *Scots Guardsman* (by then allocated to Carlisle) when it ran the 'Rebuilt Scot Commemorative Tour', which ran from Crewe to Carlisle (via the Settle and Carlisle line) on 13 February 1965. The locomotive originally chosen was 46160 *Queen Victoria's Rifleman*, but it had failed a week before with a hot box, so Crewe North prepared 46115 *Scots Guardsman* (of Carlisle Kingmoor) instead. The author noted *Scots Guardsman* on Crewe North the

week before in shabby condition, but by the time the tour was run the loco had been cleaned and a set of wooden replica nameplates fitted. Crewe continued to look after rebuilt Class 7s, with 46155 *The Lancer* receiving attention on the wheel drop in April 1965 when it was a Carlisle Kingmoor engine.

Holyhead

Holyhead was coded 7C in 1935 and recoded 6J in May 1952. It remained 6J until its closure on 12 December 1966. Holyhead was physically a small depot, with only four roads to the shed and in a cramped position, but it was an important one, as from the beginning of the rebuilding process Class 7s were allocated there. In the mid-1950s, the depot only had nineteen locomotives allocated, but six of them were rebuilt Scots.

The ferry terminal with boat services to and from Ireland generated a huge amount of passenger traffic (with over two million passengers a year travelling on the ferries) and mail traffic to Manchester, Crewe, Birmingham and London, as well as express meat trains from Holyhead to Broad Street (London) and express meat trains from Holyhead to Aston (Birmingham).

Services to and from Holyhead were marketed under the banner of 'The Irish Mail', with through trains from Manchester, Birmingham and London Euston. Another express service was given the title of 'Emerald Isle Express'. Many of these trains were very heavy and also ran overnight with sleeper cars. The parcels traffic was also extensive, with services to Manchester and Crewe, and a heavy Manchester to Bangor newspaper train was regularly a rebuilt Class 7 duty. There was a 12.00 noon parcels train from Holyhead and this was used to clear up any build-up of non-passenger stock; it could be heavy, so it was usual to seen a rebuilt Class 7 or a Duchess on this train.

During the summer, considerable extra traffic would also be generated to service the North Wales holiday resorts, as well as Holyhead for the Irish boat traffic. The intermediate stations such as Llandudno Junction would also provide passengers to travel into Snowdonia on the branch down to Blaenau Festiniog. For a short period in the early 1960s following the introduction of diesels on the West Coast Main Line,

46142 The York and Lancaster Regiment (rebuilt February 1951) is seen on the Up Welshman express on 5 September 1959. The locomotive is in excellent external condition, but the crest above the nameplate is missing. The Welshman was a regular duty for the rebuilt Class 7s based at Holyhead and Camden.

46138 The London Irish Rifleman (rebuilt June 1944) leaves Holyhead in 1960, with an express pulling a mixed bag of rolling stock and a well-filled tender of coal, which is above the cab roof. Between November 1959 and November 1962, the engine alternated between Llandudno Junction and Holyhead, before moving to Carlisle Upperby. Holyhead generated a large amount of additional summer trains and parcels traffic and was essentially the preserve of the rebuilt Class 7s until the early 1960s, when the EE Type 4s and surplus Duchesses began to take over the heavier trains. M. BENTLEY

Duchess Pacifics from Crewe and Camden could also be seen on the line, sharing duties with the rebuilt Class 7s before all the Duchesses were withdrawn in September 1964.

From early in the 1960s, a number of the key trains (such as *The Irish Mail* and *Emerald Isle*) were given over to the English Electric Type 4 (Class 40), usually from Camden, and as the years progressed more of the key trains were handed over to the diesels. Also a number of Britannias were allocated to both Holyhead and Llandudno in 1962, further reducing the regular work for the rebuilt Class 7s.

The rebuilt Class 7s were, however, still called on for the heavy trains when the EE Type 4s were not available, so in April 1963 46120 Royal Inniskilling Fusilier (6G) was observed with the 12.55 Holyhead–Euston (*The Irish Mail*) and 46156 *The South Wales Borderer* (6J) was seen working the 10.58 Euston–Holyhead at a time when most of the Euston services were in the hands of EE Type 4s.

What is clear when looking at the allocation history for the depot is the number of times engines were reallocated to the depot from Crewe North and back again. There was also a high number of loans to Holyhead, presumably to cover its own locos in Works or for extra traffic, plus the same engine could be allocated to the depot a number of times, for example 46125 *3rd Carabinier* was allocated to the depot six times between 1945 and 1963.

Llandudno Junction

Llandudno Junction was coded 7A in the LMS period, but became 6G in 1952 under 6A Chester and remained as 6G until closure on 3 October 1966.

The rebuilt Class 7s were relative latecomers to the depot, only arriving at the end of 1959 when more Class 7s were available following the introduction of the English Electric Type 4. Llandudno's engines shared duties with Bangor and Holyhead on the North Wales coast trains, with the addition of regular rains from Llandudno to Manchester, particularly the 'Club train'. This was a hangover from LNWR days, when

46152 King's Dragoon Guardsman (rebuilt August 1945) is seen on 28 June 1964 undergoing a repair at Llandudno Junction shed, probably a Piston and Valve examination (known as a 'P&V'). This was a regular part of the maintenance regime for all steam locomotives, in which the pistons and valves would be stripped down and checked (and any worn parts replaced) and excessive carbon removed at set mileages. The top lamp bracket on the smokebox has been moved to the 3 o'clock position, which would have happened sometime in 1963. The front couplings have been removed to give better access to the middle cylinder and piston valve. 46152 King's Dragoon Guardsman arrived at Llandudno Junction in August 1961, before moving to Holyhead in February 1962, where it remained until January 1965, when it moved to Carlisle Kingmoor. KEN TYLER

46166 London Rifle Brigade (rebuilt January 1945) is seen at Bangor on the North Wales coast. The engine had been allocated to a number of sheds since rebuilding, including Crewe North (fifteen times), Holyhead (six times) and Longsight (three times), before ending up at Carlisle Kingmoor from September 1963, until withdrawal in October 1964 and scrapping at West of Scotland Shipbreaking in Troon in December 1964.

46170 British Legion (rebuilt October 1935) is seen at Chester in its final condition, but with the earlier tender emblem. From 1960, it was allocated to Crewe North (four times), Holyhead and Camden, before arriving at Llandudno Junction in September 1962, whereupon it almost immediately went into storage, before being withdrawn in December 1962. It did steam for a few days as it went down to Rugby shed for the Remembrance Ceremony there on 11 November 1962, before returning to storage. The withdrawal was strange, as the engine had received a Heavy General only twelve months previously, leaving the Works in September 1961, and this repair would have included major boiler repairs. COLOUR RAIL BRM

46152 King's Dragoon Guardsman (rebuilt August 1945) is seen inside Llandudno Junction shed on 28 June 1964 on undergoing some repairs with the circular disc showing the engine was not to be moved. Llandudno Junction was the main repair depot for the sheds in the Number 6 district (covering Mold Junction, Holyhead and so on). 46152 King's Dragoon Guardsman was at the time a Holyhead engine and was a familiar sight on the North Wales coast, having been allocated to both Llandudno Junction and Holyhead for many years and the author had a trip behind the engine between Chester and Llandudno Junction in 1962. KEN TYLER

46152 The King's Dragoon Guardsman (rebuilt August 1945) is seen at Bangor station and displays its 6J Holyhead shed code. It was allocated to Holyhead for the second time in March 1962, remaining there until January 1965, when it went to 12A Carlisle Kingmoor, before being withdrawn in April 1965. The engine was scrapped at Motherwell Machinery in July 1965. The class was a regular on the North Wales line between Chester and Holyhead, with examples being seen at Holyhead and Llandudno Junction (where 46152 had been allocated between September 1961 and March 1962 before moving to Holyhead).

COLOUR-RAIL

wealthy businessmen would commute from Llandudno to Manchester and back on a daily basis and the LNWR supplied special stock consisting of saloon cars with armchairs. Whilst the stock changed over the years, the importance of this train continued up until the late 1960s. It would leave at 7.40am and was regularly loaded to eleven coaches (340tons [345,440kg]). There was also a Llandudno to Liverpool (Birkenhead) train (reversed at Chester) that was regularly loaded to eight or nine coaches.

The depot also had responsibility for through services between Llandudno and Euston (such as the 9.5am service), certainly as far as Crewe, and its locos could also be seen on the Bangor–Manchester and Portmadoc–Euston. Engines would also work the many extra trains in the summer months on the North Wales coast line to Manchester , London and Liverpool.

One of the depot's celebrity locomotives was 46170 *British Legion*, which arrived at the depot in September 1962. It almost immediately went into storage, but was steamed in November 1962 to travel down to Rugby to attend the Remembrance Ceremony held on 11 November 1962 at Rugby shed. The engine then returned to Llandudno for storage. It was withdrawn in December 1962, before being scrapped at Crewe Works in January 1963 following a period of storage there.

A feature of the depot in the early 1960s was the putting into storage at the end of the summer timetable of some of the rebuilt Class 7s. For example, 45525 *Colwyn Bay* was stored serviceable between 1 November 1962 and 3 February 1963, before being returned to traffic for a few months.

Liverpool Edge Hill

Edge Hill was coded 8A (London Midland Region) until closure. This shed was responsible for the principal trains from Liverpool Lime Street to London Euston, Liverpool to Newcastle (as far as Leeds) and Liverpool to Glasgow. It had an allocation of six Duchess Pacifics, as well as a sizable stud of rebuilt Class 7s and Jubilees. In 1959, the depot had 124 locomotives allocated, including seven Princess Royals, ten rebuilt Scots and eleven Jubilees. Many of the

rebuilt Class 7s were allocated to the depot directly from the rebuilding process at Crewe Works, with many having lengthy stays at the depot. For example, 45527 *Southport* arrived in September 1948 and did not leave until March 1961, when it went to Bushbury, and 45531 *Sir Frederick Harrison* arrived in June 1950 and did not leave until October 1963, when it moved to Springs Branch Wigan. The depot's named expresses include the '*Merseyside Express*', the '*Red Rose*' and '*The Shamrock*', the first two being regularly Pacific-hauled, while *The Shamrock* was a regular Class 7 turn.

The depot also had workings on the North and West route down to Shrewsbury, Hereford and Pontypool Road with heavy summer trains, one example being the 12.05 from Cardiff to Liverpool, which the Longsight engine would pick up at Hereford from a Western Region loco and return to Liverpool. Other workings on the North and West route would be the 9.30 Liverpool to Plymouth service.

With the dieselization of many of the services, including the trans-Pennine route (particularly the Liverpool–Newcastle services as far as Leeds), the depot lost many of its rebuilt 7 locomotives. From 2 January 1961, English Electric Type 4s (Class 40) replaced steam on all the Liverpool–Newcastle expresses and from the winter 1962 timetable the EE Type 4s were replaced by the Sulzer Type 4s (Class 46). In 1961, the depot had fifteen Class 40s on its books, with ten locos working on London diagrams and two on Newcastle diagrams (Gateshead covered most of these diagrams).

The English Electric Type 4s continued to be used on the Euston trains, with locos initially working through to Euston (refuelling at Camden). However, with the gradual extension of the electrification of the West Coast Main Line, locos could be taken off at any of the principal stations on the way to Euston. In 1961, one of the important regular steam duties remaining at the shed was as follows:

• 12.45am Liverpool Lime Street through to Glasgow Central, arriving at 7.50am with the loco being disposed at Polmadie
• 11.30pm Glasgow Central to Liverpool Lime St, arriving at 6.32am. This was the return working from the train above.

46145 The Duke of Wellington's Regiment *(West Riding) (rebuilt January 1944) is seen at Manchester London Road (now renamed Piccadilly) in late 1948 in a hybrid livery of LMS 1946 livery with a BR number on the cab side and a BR smokebox number plate. The engine was allocated to Longsight (Manchester) in April 1947 and the rebuilt Class 7s from that depot were the principal engines on the Manchester to Euston services. The engine has the early style of top feed cover and did not receive smoke deflectors until January 1951.*

To cover the diagram two locos were allocated, one loco working alternatively to allow for servicing, inspections and failures.

In the late 1950s and early 1960s, a complex diagram had 8A Class 7 locos working an afternoon stopping train from Stafford to Shrewsbury, which then allowed the loco to work forwards from Shrewsbury to Crewe and Liverpool. So in 1958, an Edge Hill rebuilt Class 7 loco would work the 8.55 Stafford – Shrewsbury and the 11.25 back to Stafford, before the loco continued to Birmingham. The 3.10pm Shrewsbury–Stafford was also worked by an Edge Hill Class 7, which arrived at Shrewsbury on the 11.45 Manchester (London Road)–Plymouth (at Shrewsbury the train would be handed over to a Western Region loco) and then worked the local to Stafford and back to Shrewsbury, where the engine could pick up another express back to Crewe. This enabled the author in 1962 to see a number of 8A locos in a short space of time, including 45531 *Sir Frederick Harrison*, 45535 *Sir Herbert Walker*, 46119 *Lancashire Fusilier* and 46134 *Cheshire Regiment*.

However, in the event of non-availability of the scheduled diesel locomotive, the depot had to 'borrow' any available locomotives, so as an example on 6 October 1961 46119 *Lancashire Fusilier* was seen deputizing for a diesel on the 3.16pm Newcastle–Liverpool from Leeds, but it slipped to a stand on the 1 in 70 incline through Gildersome Tunnel and had to be banked up the rest of the incline.

On 11 November 1963, 46168 *The Girl Guide* (Springs Branch Wigan) was appropriated for the 9.00am Liverpool–Newcastle service, where it only dropped three minutes on the diesel timings as far as Huddersfield. The locos were increasingly used near the end of their time at the depot on specials and holiday extras, as on 25 January 1964 when 45531 *Sir Frederick Harrison* was used on a heavy (approximately ten coaches) Liverpool–Leeds football excursion.

London Camden

Camden was the primary shed for London Euston and was coded 1B. It had an extensive allocation of rebuilt Class 7s, as well as Princess and Duchess Pacifics. Camden's diagrams covered all types of expresses, including trains to Crewe, Holyhead, Blackpool, Preston, Wolverhampton and even the Royal Ordnance Factory trains between Chorley and Wigan.

Camden was also responsible for some of the turns from Holyhead, so the 4.00pm Holyhead–Euston was a 1B locomotive, usually a rebuilt Scot, but in the early 1960s Camden would provide a Duchess or an EE Type 4 (Class 40).

46139 The Welch Regiment (rebuilt November 1946) is seen in the late 1950s, when it was allocated to 1B Camden (from June 1947) until its transfer to 14B Kentish Town in November 1959. It then moved to 26A Newton Heath in September 1961. The engine was withdrawn in October 1962, before being scrapped at Crewe Works in May 1963. The regimental crest above the nameplate is missing and the AWS has not yet been fitted (on 18 April 1959); neither has the speedometer drive (fitted 21 May 1960). The engine still has its original Fowler driving wheels, but the bogie wheels are now Stanier with bevel rims.

After the two rebuilt Jubilees, 5735 *Comet* and 5736 *Phoenix*, had proved their worth, they were reallocated to Camden in July 1943 and were closely followed by a number of rebuilt Patriots and rebuilt Scots as they exited Crewe from the rebuild process. The first Jubilee observed working was 5735 *Comet* on 18 August 1943, with a fourteen-coach train on the Euston–Heysham run. As an example of one of Camden's diagrams, Turn 12 was as follows:

- 11.39pm Preston–Euston (6.28 from Workington)
- 10.40am Euston–Blackpool Central
- 8.00am Blackpool Central–Euston
- 10.50pm Euston–Bletchley
- 12.25am Bletchley–Preston parcels (9.40 from Marylebone)
- 4.52pm Euxton (Royal Ordnance Factory)–Wigan North Western
- 6.33am Wigan North Western–Euxton (ROF).

With the allocation of English Electric Type 4s (Class 40) and a change to the diagramming, responsibility for some former Camden workings was passed to Pre-ston (including the diagram above) and a small number of rebuilt Class 7s were transferred to Preston shed in September 1959 to cover these services. With the closure of Preston in September 1962 due to fire damage, these were then moved to 8F Springs Branch. During 1959, a number of Camden's rebuilt Class 7s were transferred away to the Midland division, mainly to 14B Kentish Town, as much of their former work had been taken over by EE Type 4s.

By November 1961, the depot had twenty-three EE Type 4s allocated to it, reducing the need for its remaining steam locomotives, which were transferred to 1A Willesden. Camden was gradually run down and closed in September 1962.

London Willesden

Willesden was allocated Code 1A until closure. It was essentially a freight depot, as it was adjacent to the extensive freight yard at Willesden, but it also had responsibility for some fast freights that would require 4-6-0 power rather than an 8F 2-8-0. In 1948, the

46135 The East Lancashire Regiment (rebuilt January 1947) is seen in August 1962 at Willesden shed. The engine is fitted with both AWS (January 1960) and speedometer drive (September 1961). It also has a replacement set of leading driving wheels that have Stanier-style balance weights, whilst the balance weights on the centre axle have been modified from the usual Fowler pattern. The bogie wheels are now of the Stanier type. The engine is in the careworn condition common to the rebuilt Class 7s at the time and had been allocated to 5A Crewe North in September 1960, before moving to 55A Leeds Holbeck in August 1962. Withdrawal took place in December 1962. The engine was then stored at Crewe Works between December 1962 and March 1963, before being scrapped. COLOUR-RAIL

45530 Sir Frank Ree (rebuilt October 1946) is seen at Willesden in April 1963 and seems to have been the depot 'pet', as it was always kept in excellent condition, having arrived at Willesden for the second time in July 1962. The engine subsequently went into Crewe Works for an overhaul in August 1963. Following a move to Holyhead in June 1964, it went to Carlisle Kingmoor in January 1965 and was withdrawn from Kingmoor in December 1965. The engine has the later style of Stanier wheels (drivers and bogie) with a bevelled rim and is fitted with AWS (March 1959) and a speedometer drive (May 1962). It is fitted with the welded tender it acquired at rebuilding. COLOUR-RAIL

45736 Phoenix (rebuilt April 1942) is seen at Willesden shed in October 1963, to which the engine had been allocated in June 1963, arriving from Holyhead. The engine remained at Willesden, before moving to Carlisle Upperby in July 1964, then on to Kingmoor in September 1964. The engine has AWS (September 1959) and a speedometer (May 1961). It has the later style of top feed cover, plus a welded tender. Looking at the rust forming on the slide bars and piston rod, the engine has been out of service for a few weeks. COLOUR-RAIL

46110 Grenadier Guardsman (rebuilt January 1953) sits in Willesden shed on 1 July 1962, displaying its 8A Edge Hill shed plate. At this time, Willesden was the usual depot for London steam, as Camden was being run down and would close in September 1962. As a result, Willesden had already acquired most of Camden's rebuilt Class 7s. The locomotive was to survive until February 1964, being withdrawn from Carlisle Kingmoor, and remained in store at Carlisle Upperby until November 1964, when it was scrapped at McWilliams at Shettlestone near Glasgow. RAILONLINE

shed's allocation was 126 steam locomotives, which included three Patriots, five Jubilees and eight Black Fives. The depot had four Class 6 turns for its Jubilees and Patriots, including an SO Bletchley–Blackpool, a Crewe–Liverpool–London passenger and an FO Euston–Holyhead passenger train. A newspaper train to Northampton and return with a passenger train was also a Willesden responsibility. The top freight working for the depot was a Class C freight to Glasgow, which was the fastest freight train in Great Britain, and there were extensive parcels workings for the depot.

The depot also had extensive repair facilities and did not suffer from a chronic lack of space, as happened at Camden. It was common practice for Willesden to carry out regular maintenance on Camden's fleet of passenger engines, including the rebuilt Class 7s and Pacifics.

In 1961, the majority of Camden's rebuilt Class 7s were reallocated to Willesden, although Willesden had also had a small number of Jubilees and rebuilt Class 7s prior to the closing of Camden.

Some of the remaining express duties from Camden were transferred to Willesden and in the event of diesel non-availability, Willesden would supply a rebuilt Class 7 or one of the Duchesses it inherited from Camden. Also, many extra summer trains or excursions would see Willesden's rebuilds on the main line. The fast outer suburban trains out of Euston could also call on rebuilt Class 7 power, so on 19 May 1964 the Euston–Rugby train was seen with Willesden's pet loco, 45530 Sir Frank Ree, at its head. Football and hockey specials to Wembley made good use of Willesden's rebuilt Class 7s and they were also regularly seen on Holyhead services, so in 1962 46126 Royal Army Service Corps was seen on 'The Welshman', 45530 Sir Frank Ree was seen on a Holyhead–Euston service and 45736 Comet was also spotted on a Holyhead–Euston extra.

Also, the dieselization of most of the services from Holyhead to Euston and Birmingham led to an influx of rebuilt Class 7s, arriving from Holyhead in June 1963. These included 45527 Southport, 45736 Phoenix, 46114 Coldstream Guardsman, 46125 The 3rd Carabinier, 46150 The Life Guardsman and 46156 The South Wales Borderer.

A feature of the rebuilt Class 7s allocated to Willesden was the use of Devons Road Bow for the storage

45529 Stephenson (rebuilt July 1947) sits at Willesden shed alongside one of the original Patriots. The engine is still allocated to 5A Crewe North, but would soon be reallocated to Willesden in January 1961. It then moved to Annesley in October 1963, before being withdrawn February 1964. COLOUR-RAIL

45528 REME (rebuilt August 1947) is seen on 28 April 1962 in front of the old LNWR repair shop at the back of Willesden shed with the rear wheel set of the tender missing. REME had arrived at Willesden in January 1961 and remained there until withdrawal in January 1963. After a period stored at Crewe Works, it was cut up there in March 1963. The engine had been unnamed when rebuilt and only received its name on 2 October 1959 without any naming ceremony. The engine has AWS (January 1960) and a speedometer drive (August 1961) and like most of the rebuilt Patriots has a welded tender. RAILONLINE

of engines, particularly over the winter period, and photos show four or five locos stored in a line at the depot, some being reinstated to traffic, but some being withdrawn.

Most of the rebuilt Class 7s allocated to 16B Annesley were transferred from Willesden, so it was not sur prising the depot did not send its best engines and some of these engines were seen regularly at Willesden after transfer away, as it was also responsible for providing servicing facilities (along with Cricklewood) to the rebuilt Class 7s working on the ex-Great Central Railway (GCR) route into Marylebone, as Neasden had been closed in June 1962.

With the introduction of the EE Type 4s at Camden and changes to the diagrams, Willesden engines could be seen regularly on the Euston–Wolverhampton trains (as well as EE Type 4s and Princess Coronation Pacifics), so as an example in April 1961, 46120 *Royal Inniskilling Fusilier* was seen on the Wolverhampton–Euston service and 46156 *The South Wales Borderer* was seen working the 9.20 Euston–Wolverhampton train on a number of occasions in 1963.

Willesden's rebuilt Class 7s were also seen on many of the North Wales services to Holyhead, as well as on the Grand National special in March 1962 and on the Euston–Blackpool Illuminations specials. The shed closed in September 1965.

Glasgow Polmadie

Polamadie was in the Scottish Region and bore the code 66A. Its principal workings (along with the depot's Duchess Pacifics) were from Glasgow Central to Carlisle and further south to Crewe, as well as passenger trains to Edinburgh Princess Street. The Perth–London Euston expresses were also a Polmadie rebuilt Scot turn. Generally, Polmadie's rebuilt 7s would come off southbound trains at Carlisle, but some of them would work to Manchester Victoria, sharing the workings with Newton Heath Jubilees. In a nice touch, the shed's allocation of rebuilt Scots was all named after Scottish Regiments.

46118 Royal Welch Fusilier *(rebuilt December 1946) is seen at Polmadie, Glasgow, in September 1956. At the time of the photograph, the engine was based at Crewe North, having arrived in September 1954 and departing for Nottingham in January 1960. The engine was to receive AWS November 1959 and a speedometer in October 1960. It ended its days at Carlisle Upperby in June 1964. As one of the oldest Regiments in the British Army (being raised in 1689), the Regiment was allowed to continue with the old spelling of Welch. The driving wheels are the original Fowler-style wheels, whilst the bogies have been changed for Stanier wheels.*
COLOUR-RAIL

Rebuilt Class 7s at Polmadie

Loco	From	Date	To	Date	Comments
46102 *Black Watch*	Crewe Works	October 1949	Corkerhill	October 1962	Allocated to Polmadie when in original condition. After rebuilding, went back to Polmadie. Stored at Corkerhill, December 1962 to May 1964.
46104 *Scottish Borderer*	Crewe Works	March 1946	Corkerhill	October 1962	Allocated to Polmadie when in original condition. After rebuilding, went back to Polmadie. Stored at Corkerhill, December 1962 to May 1964.
46105 *Cameron Highlander*	Crewe Works	April 1948	Withdrawn	December 1962	Allocated to Polmadie when in original condition. After rebuilding, went back to Polmadie. Stored at Polmadie, December 1962 to May 1963 and then Parkhead (65C), May 1963 to May 1964.
46107 *Argyll and Sutherland Highlander*	Crewe Works	February 1950	Withdrawn	December 1962	Allocated to Polmadie when in original condition. After rebuilding, went back to Polmadie. Stored at Polmadie, October 1962 to May 1963 and then Parkhead (65C), May 1963 to May 1964.
46121 *Highland Light Infantry*	Longsight	July 1949 (initially on loan, then permanent)	Withdrawn	December 1962	Stored at Polmadie, October 1962 to May 1963 and then Parkhead (65C), May 1963 to May 1964.

46107 Argyle and Sutherland Highlander (rebuilt February 1950) is seen at its home depot of 66A Polmadie in April 1962. The engine spent its entire life as a rebuild at Polmadie, being withdrawn from the depot in December 1962 when all the rebuilds there were withdrawn. The engine then spent many months in storage, first at 66A Polmadie and next at 65C Parkhead. It was scrapped at McWilliams, Shettlestone, near Glasgow in May 1964. COLOUR-RAIL

One of Polmadie's regular duties for its Scots was the 11.15 Glasgow–Birmingham as far as Crewe, but the locomotive's crews were from the Glasgow link at Crewe North shed. Fireman Peter Johnson reported that all of the Polmadie Scots were very rough riders and that, despite firing it many times, he could not get 46104 *Scottish Borderer* to steam. He compared the riding qualities of the Polmadie engines to the ones he fired from Leeds Holbeck, which in his words 'steamed and rode like a coach', despite being long out of Works.

On 29 January 1962, 46102 *Black Watch* was seen acting as station pilot at Huddersfield, although the author has been unable to find out what the background to this was. This engine was also seen in August 1962 pulling a train of hopper wagons – not exactly an express passenger working, which seems to indicate that even in the summer of 1962 much of the express work had already been taken away by the new diesels.

The severe winter of 1962–3 caused problems for Polmadie, as it had withdrawn all its rebuilt Class 7s at the end of 1962 and with the non-availability of some of its diesels, the depot had to borrow 46166 *London Rifle Brigade* from Crewe North for a few days to cover some of its diagrams in early 1963. Upon withdrawal, the Polmadie Scots spent almost two years in storage, before finally being scrapped in the Glasgow area at McWilliams at Shettleson.

Glasgow Corkerhill

Corkerhill carried the shed code 67A. It was the ex-Glasgow and South Western passenger depot for Glasgow St Enoch station. Its usual allocation of 4-6-0s consisted of a stud of Black Fives and Jubilees, essentially for services to Kilmarnock, Ayr, Stranraer and Carlisle. In 1960, the depot had eleven Jubilees on its books (45621, 45665, 45673, 45677, 45687, 45693, 45692, 45707, 45711, 45720 and 45727), but by October 1962 all of them had been withdrawn, which coincides with the arrival of the two rebuilt Scots, so it may have been a decision that the two rebuilt Scots could provide some stand-by steam power in the event of diesel non-availability.

Rebuilt Scots had been common visitors to the depot, as Leeds Holbeck engines would be serviced there after working on the Leeds–Glasgow St Enoch trains, but none was allocated there until the end of 1962. However, they only remained operational for two months before withdrawal, so it is not known if they actually carried out any work when at Corkerhill. They certainly spent a long time in storage after withdrawal, as they were there for eighteen months before final scrapping. Rebuilt Scots continued to be seen at the shed, however, with 46162 *Queen's Westminster Rifleman* (12A Carlisle Kingmoor) being seen at the depot on 29 March 1964.

Rebuilt Class 7s at Corkerhill

Loco	From	Date	To	Date	Comments
46102 *Black Watch*	Polmadie	October 1962	Withdrawal	December 1962	Stored at Corkerhill, December 1962 to May 1964. Scrapped at McWilliams yard at Shettlestone between May 1964 and July 1964.
46104 *Scottish Borderer*	Polmadie	October 1962	Withdrawal	December 1962	Stored at Corkerhill, December 1962 to May 1964. Scrapped at McWilliams yard at Shettlestone between May 1964 and July 1964.

46102 Black Watch *(rebuilt October 1949) is seen on 23 June 1963 at Corkerhill, where the engine had been allocated just before it was withdrawn at the end of December 1962. The engine had been a long-term resident of Polmadie depot, spending almost its entire career as a rebuilt Class 7 there until the late transfer to Corkerhill. The engine remained in store at Corkerhill for eighteen months and did not leave until May 1964, when it went to McWilliams at Shettlestone. It lingered there for another couple of months, before finally being scrapped in July 1964. The engine has been fitted with both AWS and a speedometer drive at some point around 1960 or 1961.* DAVID COUSINS

Manchester Longsight

Longsight, coded 9A, was the ex-LNWR shed that provided locomotives for the routes from Manchester London Road (later renamed Piccadilly) to Stoke-on-Trent, Crew and Euston, as well as Stockport. The Euston expresses could be routed via Crewe or Stoke. The principal turns for Class 7s were expresses to London via Crewe and Stoke from Manchester London Road, with a number of the trains being given titles such as *'The Mancunian'* and *'The Lancastrian'* and during the summer the *'Pines Express'* from Manchester London Road to Bournemouth. Longsight also had workings down to Shrewsbury and Hereford on summer services from Manchester to Penzance services. One of the diagrams would be the 4.45pm Penzance–Manchester service, with the Longsight rebuilt Scot coming on to the train at Hereford. The engine had arrived at Hereford on the 4.15 Coleham

(Shrewsbury) to Hereford goods, where it was then serviced and coaled (with coal sent to the depot specifically for the service) at Hereford shed and waited for the incoming train from Penzance.

There was also a Longsight working with the 9.25 Manchester Swansea as far as Pontypool Road, returning some two hours later with the 7.30 Penzance–Manchester service, which could be worked by either a Longsight Jubilee or a rebuilt Class 7. There was a short turnaround at Pontypool Road, as the loco would arrive at 1.25pm and would need to be out of the shed at 3.00pm ready to pick up the train at 3.19pm, but at least the engine would have received a tender of good Welsh steam coal.

Longsight became the home of the first rebuild when 6170 *British Legion* was allocated there in November 1935. It was put straight on to top-link working, being regularly seen on the up *'Mancunian'* and the down *'Lancastrian'*. During its stay, it also

46155 The Lancer (rebuilt August 1950) is seen at Buxton shed on 15 July 1964 and had been rostered to work the 8.25pm Buxton–Manchester Mayfield parcels train, having worked into Buxton on the morning parcels train from Mayfield. The engine at the time was allocated to Crewe North, leaving for Carlisle Kingmoor in September 1964. The AWS air tank is visible (fitted July 1959) and the speedometer drive (fitted July 1961). The engine's final allocation was to Carlisle Kingmoor in September 1964, until withdrawal in November 1964. It was scrapped at West of Scotland Shipbreaking, Troon, in February 1965. Unusually, the locomotive still has its nameplates, but when photographed in late 1963/early 1964 the nameplates had been removed, so the assumption is that the original plates have been restored. KEN TYLER

45522 Prestatyn (rebuilt February 1949) is seen at its home depot of Longsight (Manchester) on 22 September 1963, but without a shed plate. The engine had arrived at the depot in June 1963 from Newton Heath and remained at Longsight until withdrawal in September 1964, although it had been borrowed by Buxton for use on the Manchester Central to Buxton fast trains and was actually withdrawn at Buxton and remained there stored until May 1965. Buxton seemed to take more care of the loco, as it was observed in good external condition when working the Buxton trains. RAILONLINE

46129 Scottish Horse (rebuilt December 1944) is seen at Cheadle Heath station on 26 March 1964 on the 5.22pm Manchester Central–Buxton commuter train. The engine (a Longsight resident) was a regular on this train (and the corresponding 8.00am morning train from Buxton) in the winter of 1963–4, with its last observed working being 8 May 1964. At the Manchester end, the engine would stand at Trafford Park shed 'spare' during the day and similarly at the Buxton end the engine would remain on shed until working the train the following day. The engine would also sit at Buxton shed over the weekend, ready for the Monday morning train. It was allocated to Longsight in September 1962 and was withdrawn in June 1964, being stored at Longsight between June 1964 and October 1964, before being towed to Central Wagon Co. at Ince, near Wigan, in November 1964 for scrapping. KEN TYLER

45522 Prestatyn (rebuilt February 1949) is seen at the side of Buxton shed shortly after withdrawal in September 1964. The engine was allocated to Longsight, but seems to have been borrowed by Buxton and was regularly seen on the 8.00am Buxton–Manchester Central and returning on the 5.22pm Manchester Central–Buxton. 45522 Prestatyn was observed on this daily working from 14 June 1964 until 18 September 1964. The AWS (November 1959) and speedometer drive (February 1961) can be seen, as well as the yellow warning stripe applied at the end of August 1964. This was only applied between the cab lining rather than across the full cab side. The nameplates are still in situ, which was unusual for a Class 7 rebuild as most had lost their plates in 1963. The engine remained stored at Buxton between September 1964 and May 1965 and in the process its double chimney disappeared. Following storage at Buxton, the engine departed to Central Wagon Co. at Ince, near Wigan, where it was scrapped in June 1965. KEN TYLER

46115 Scots Guardsman (rebuilt August 1947) is seen at Buxton shed in May 1964, where for four weeks it was the regular engine on the 8.00am Buxton–Manchester Central and the return service, which was the 5.22pm Manchester Central–Buxton. By this date, the speedometer drive had been removed, just leaving the operating crank on the rear axle. The engine was based at Manchester Longsight, before moving to Carlisle Kingmoor. It had become the last operational rebuilt Class 7 when it was withdrawn in December 1965.
KEN TYLER

became a 'film star' appearing in the title role of the 1936 LMS film *Engine on the Shed*, which can be thoroughly recommended.

In 1950, the shed had thirteen rebuilt Class 7s allocated. By 30 January 1954, it had thirteen rebuilt Scots and four rebuilt Patriots on its books. As well as the regular express workings from Manchester, the rebuilt Class 7s would be used for specials, so for the 1953 Football Cup Final the newly rebuilt 46140 *The King's Royal Rifle Corps* was used on one of the Cup Final specials and for the Rugby League Cup Final the same year that three of Longsight's rebuilt Scots, 46114 *Coldstream Guardsman*, 46115 *Scots Guardsman* and 46169 *The Boy Scout*, put in an appearance. Longsight also worked a 9.10 Manchester–Swansea down the North and West route as far as Pontypool Road; during the summer this was a heavy load, with thirteen coaches not being unusual. During the winter, the load was more usually seven coaches.

Longsight became responsible for some of the diagrams formally the responsibility of Trafford Park (which was a sub-shed of Longsight). For example, the rebuilt Class 7 working to Manchester Central from Buxton in the morning and the return working in the evening would leave the rebuilt Class 7 as a 'spare' engine at Trafford Park, where it could act as a stand-by engine in the event of diesel failure. The rebuilds could also then be used on summer Saturday specials and 45522 *Prestatyn* was recorded on a Buxton–Windermere and return working.

In November 1961, Longsight had fourteen EE Type 4s allocated to it for its principal passenger workings. In the declining years at Longsight, the rebuilt Class 7s were rostered for the important Buxton–Manchester and return commuter train. 46129 *The Scottish Horse* was first rostered, but it is believed that the engine worked it through the winter of 1963–4, with the last recorded working on 8 April 1964. 46115 *Scots Guardsman*'s first recorded working of the service was 3 May 1964, with its last recorded working being 7 June 1964. The first recorded working of 45522 *Prestatyn* was 14 June 1964, with the last recorded working being 20 September 1964. The engine was retubed at Buxton immediately before withdrawal, another example of wasted costs in the rundown of steam. The shed closed to steam on 14 February 1965.

Wolverhampton Bushbury

Bushbury's shed code was 3B, which changed to 21C in July 1960. It was the ex-LNWR/LMS shed just north of Wolverhampton High Level station (the ex-LMS one to differentiate it with the Low Level, which was the ex-GWR station). Its principal passenger workings were the Wolverhampton–Euston expresses via Birmingham. Whilst Bushbury provided the majority of the motive power for these services, some of the diagrams were covered by Camden shed. Prior to World War II, these expresses were in the hands of non-rebuilt Patriots, but after the war they received a number of the rebuilt versions.

As they were rebuilt and emerged from Crewe Works (or very shortly afterwards; a number seem to have spent their first few weeks on loan before arriving at Bushbury, most of the rebuilt Patriots were allocated to Bushbury to haul the Wolverhampton, Birmingham–Euston trains. Some of these rebuilt locomotives were familiar to the Bushbury crews, as they had been allocated to the depot before being rebuilt (5512 *Bunsen*, 5525 *Colwyn Bay*, 5526 *Morecambe and Heysham* and 5531 *Sir Frederick Harrison*). They were well received by the Bushbury crews, with one exception – they did not like the smoke drifting down the side of the boiler. The double chimney and blast pipes produced a much softer exhaust compared to the non-rebuilt versions and smoke would drift down the boiler sides, obscuring the view ahead and it resulted in complaints from the drivers. The non-rebuilt Patriots also had smoke deflectors to improve the visibility ahead. Keith Terry (a Bushbury fireman) recounts the tale of Jack Salter (the most senior driver at the depot) and some of his more senior colleagues officially complaining over the issue and as a result the rebuilt Patriots were transferred away – all had gone by July 1950. The replacements were Jubilees, so by the end of September 1950 the shed had eight Jubilees on its books. The fitting of smoke deflectors to the rebuilt Patriots did not commence until the end of 1950 and was not completed until 1953, long after the complaints from the senior Bushbury drivers.

The service was upgraded for the autumn of 1950 and the loading increased to 400tons (406,400kg). A number of original Royal Scots (which were fitted

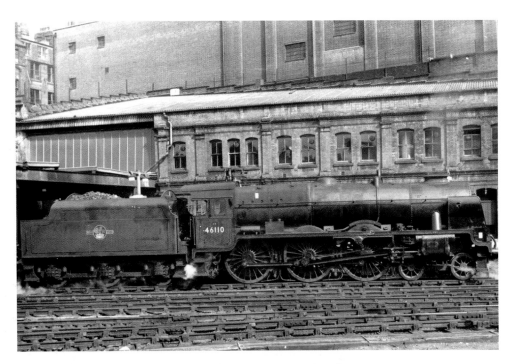

46110 Grenadier Guardsman (rebuilt January 1953) is seen at Birmingham New Street. The photograph was taken sometime after April 1959, when the engine was fitted with AWS and has also acquired the overhead warning flashes sometime in 1960. The rebuilt Class 7s were regular visitors at New Street, with trains from Manchester and Liverpool via Birmingham–Euston, as well as the Wolverhampton–Euston expresses. Between October 1958 and November 1961 the engine was on Crewe North's books, before moving to Edge Hill in November 1961, where it remained until June 1963 when it moved to Carlisle Kingmoor.

45540 Sir Robert Turnbull (November 1947) is seen at Birmingham New Street in 1948 in the experimental 'light green' livery in which it was painted in May 1948. The livery was short-lived, as BR decided not to proceed with this paint scheme, instead adopting the LNWR style of lined black for mixed traffic engines and green for express locomotives. Another rebuilt Patriot, 45531 Sir Frederick Harrison, was painted in the same livery, as was one rebuilt Scot, 46139 The Welch Regiment. The engine was allocated to 3B Bushbury upon rebuilding and is probably on a Wolverhampton–Euston train.

with smoke deflectors) were allocated to the depot. These were 46110/34/40/48/51/58/63/65, with 46140 *The King's Royal Rifle Corps* working the inaugural '*Midlander*' express on 25 September 1950. However, the stay of the original Royal Scots was short and by July 1950 they had been transferred away, the trains again reverting to Jubilees.

In November 1959, the majority of the Euston–Birmingham–Wolverhampton services were withdrawn and the services on the Wolverhampton–Birmingham Snow Hill–Paddington were augmented, the reason being the civil engineering work for the rebuilding of Coventry station and other engineering work on the LM route prior to electrification. The effect was to reduce the remaining Euston–Birmingham–Wolverhampton trains to four each way and for this reduced service Bushbury received five rebuilt Royal Scots and one rebuilt Patriot:

- 45540 *Sir Robert Turnbull*
- 46122 *Royal Ulster Rifleman*
- 46141 *The North Staffordshire Regiment*
- 46143 *The South Staffordshire Regiment*
- 46153 *The Royal Dragoon*
- 46158 *The Loyal Regiment*.

A nice touch was the allocation of both 46141 *The North Staffordshire Regiment* and 46143 *The South Staffordshire Regiment*, as both Regiments had local associations and recruited in the area.

Whilst Bushbury was primarily responsible for most of the express workings, the balance of the workings became the responsibility of Crewe North, which usually provided a rebuilt Class 7, but by 1962 it was not unusual to see a Princess Coronation Pacific on these workings.

Two more rebuilt 7s, 46114 *Coldstream Guardsman* and 46161 *King's Own*, were allocated in April 1961, but all the rebuilds had gone by June 1961. The reduced workings were also covered by locos from Camden and increasingly these were English Electric Type 4 diesels (Class 40) rather than steam locomotives (the author's first sight of a Class 40, D233 *Apapa*, was on one of these services in 1962). By 1961, the amount of engineering work for electrification was increasing and the alternative Western

Region service was further accelerated, with the introduction of the Western diesels in September 1962.

Rebuilt Class 7s could still put in appearances on express passenger trains, however, so 46156 *The South Wales Borderer*, which was a recent transfer to Willesden in June 1963, was seen working the 9.20pm Euston–Wolverhampton on a number of occasions in 1963.

Carlisle Kingmoor

Kingmoor was coded 68A (Scottish Region) until 23 February 1958, then 12A (London Midland Region) from 23 February 1958 until closure. The shed was located to the north of Carlisle and was previously the Caledonian Railway depot. Many trains changed locomotives at Carlisle, both north and southbound, as well as some trains terminating in the city. Kingmoor engines generally worked services to Glasgow and to Leeds and were also regularly seen on services up to Perth. Up until 1961, rebuilt Class 7s were virtually unknown at the depot, with the express passenger locomotives being the Duchess Pacifics and Jubilee 4-6-0s. However, from 1961 onwards the depot became a bit of a dumping ground for rebuilt Class 7s from all over the London Midland Region.

In later years, the Aberdeen–London fish train was a regular rebuilt Class 7 working, with a Kingmoor engine coming on to the train at Perth. In 1964, 46160 *Queen Victoria's Rifleman* was working the train from Perth and also in July 1964 45527 *Southport* was seen at Perth, with the engine coming off the train at Carlisle.

During the extreme winter of 1962–3, Kingmoor depot had its rebuilt Class 7s borrowed by other depots to cover the non-availability of diesels, with 46116 *Irish Guardsman* seen on the Manchester–Newcastle service on 29 March 1963; Bristol Barrow Road borrowed 46128 *The Lovat Scouts* for the depot's diesel standby on 1 January 1963.

The depot had a wheel drop and this was still being used for maintenance on rebuilt Class 7s in the mid-1960s, with 46128 *The Lovat Scouts* observed with its leading drivers removed in 1964 and 46115 *Scottish Guardsman* with its centre driving wheels removed in 1965.

45531 Sir Frederick Harrison (rebuilt December 1947) is seen at Carlisle Kingmoor on 9 August 1965. It was one of seven rebuilt Class 7s to survive into 1965 and when withdrawn in November 1965 was one of final four. The tender is the riveted version, which was not usual on the rebuilt Patriots and the nameplates have finally been removed, although they were still in place in August 1964 when the locomotive acquired its yellow warning stripe. PETER GROOM

46140 The King's Royal Rifle Corps (rebuilt May 1952) was one of the last survivors, not being withdrawn until November 1965. 46140 is seen arriving at Carlisle Kingmoor yard with a short freight. By 1964–5, the duties of the rebuilt Class 7s were extremely varied, with freight working common. By this time, Kingmoor had a large allocation of Britannias, further reducing the passenger work for the rebuilds. G. SHARPE

The depot became famous for the fabulous cleaning jobs its staff did when rebuilt Class 7s were used on football specials, so in January 1964 45736 *Phoenix* and 45527 *Southport* were given a good clean to pull FA Cup specials, involving Carlisle United playing Luton, from Carlisle to Leeds, where the locos came off the train and it continued to Luton.

On 18 May 1964, 45527 *Southport* was loaned to Blackpool shed and worked an eleven-coach Preston–York excursion, arriving at Skipton before time. The crew declined to wait for the booked assisting engine (which was still on shed) and worked the steep, sharply curved line out of Skipton without a slip, a tribute to both the crew from Bradford Manningham shed (who had crewed the loco from Colne) and to the loco.

Kingmoor was responsible for providing power for the 17.38 Manchester–York express and on 7 May 1965 45531 *Sir Frederick Harrison* was observed on the train at Huddersfield. The loco returned to Manchester the following day on the Newcastle–Red Bank empty vans, double-heading with a Black Five.

By the end of 1964, Kingmoor was the last shed to have rebuilt Class 7's allocated, consisting of 45512 *Bunsen*, 45530 *Sir Frank Ree*, 45531 *Sir Frederick Harrison*, 46115 *Scots Guardsman*, 46128 *The Lovat Scouts*, 46140 *The King's Royal Rifle Corps*, 46152 *The King's*

Dragoon Guardsman and 46160 *Queen Victoria's Rifleman*. There was little regular passenger work for them, so the depot seems to have used them on any sort of work, with 46152 *The King's Dragoon Guardsman* being seen at Leicester on 22 January 1965, having worked on a freight from Carlisle, and 46140 *The King's Royal Rifle Corps* was seen regularly on the West Coast Main Line on express freights and was seen at Polmadie on 23 August 1965. Some of the trains were even less glamorous, with 46115 *Scots Guardsman* being seen on the Settle and Carlisle route with a rake of 16ton (16,256kg) mineral wagons. However, on 1 April 1965 45530 *Sir Frank Ree* was seen on the 1.26 Carlisle–Perth express, which would have normally been a Carlisle Britannia turn.

46115 *Scots Guardsman* was regularly diagrammed for the 15.40 Bradford Foster Square–Carlisle slow passenger and 46140 *King's Royal Rifle Corps* was seen on 17 September 1965 on a Leeds–Blackpool illuminations special. On 24 March 1965, 46160 *Queen Victoria's Rifleman* was seen double-headed with a Stanier Black Five on the 11.00 Liverpool–Manchester express. During the course of 1965, the survivors were gradually withdrawn, so by the end of May 1965 the only two remaining locomotives were 46115 *Scots Guardsman* and 46160 *Queen Victoria's Rifleman*. 46140

45512 Bunsen (rebuilt July 1948) is seen at Carlisle Kingmoor shed on 19 August 1964 and displays the newly applied yellow warning stripe that had to be in place by 1 September 1964. 45512 had been a long-time resident of Carlisle, having been allocated to Carlisle Upperby in May 1949, where it stayed until it moved across the city to Kingmoor in November 1964. It remained at Kingmoor until withdrawal in March 1965, before being scrapped at Motherwell Machinery and Scrap, Wishaw, in July 1965. The engine has the welded tender, which it acquired in 1952, and has AWS (November 1959) and speedometer (February 1961). KEN TYLER

46132 The King's Regiment Liverpool (rebuilt November 1943) is seen in Carlisle. It was allocated to Carlisle Upperby in June 1962, then moved across the city to Carlisle Kingmoor in September 1963, before being withdrawn in February 1964.
G. SHARPE

was withdrawn in October, with 46115 hanging on until New Year's Day 1966, having been seen in Blackpool as late as September 1965 pulling a Glasgow Holiday special.

As can be seen by the listed allocation, Kingmoor became the last depot for most of the rebuilt Class 7s allocated to it, with most being withdrawn from the depot and disposed of at private scrapyards in the south of Scotland.

Carlisle Upperby

Upperby was coded 12A (London Midland Region) until 23 February 1958, then 12B (London Midland Region) from 23 February 1958 until closure. The shed was located to the south of Carlisle station and was previously the LNWR depot for the city. In BR days, it had a large allocation of Stanier Pacifics, as well as rebuilt Class 7 locomotives and Jubilees. The depot essentially worked on services south of Carlisle on the ex-LNWR Main Line and many Glasgow–Euston trains would change engines at

Carlisle, so the southbound journey would be the responsibility of Upperby. The northbound trains would again have the engine changed at Carlisle. with Kingmoor or Glasgow Polmadie engines taking the train northwards.

Upperby also had a diagram for an express freight to Manchester and this service was regularly worked by rebuilt Class 7s, with the engines being serviced at Patricroft depot before making the return journey.

In November 1961, eight EE Type 4s were allocated to the depot, which started to make inroads into the steam workings from the depot. In November 1962, 46106 *Gordon Highlander*, 46127 *Old Contemptibles*, 46137 *The Prince of Wales's Volunteers* (South Lancashire), 46141 *The North Staffordshire Regiment* and 46166 *London Rifle Brigade* were all in store at Carlisle Upperby, but many were to be taken out of store at the end of the year due to the extreme weather conditions, including two engines that had been withdrawn. Both of the Carlisle depots' rebuilt Class 7s were borrowed by other depots to cover for diesel non-availability. As an example, in January 1963 the Liverpool–Newcastle service was pulled by rebuilt Class 7s on many occasions, with 45545 *Planet*, one of Upperby's rebuilt Patriots, doing the work on 10 January.

On 16 November 1963, 46118 *Royal Welch Fusilier* worked a heavy Carlisle to York football special and was

45545 Planet (rebuilt November 1948) is seen in store at Carlisle Upperby depot on 19 August 1964, having been withdrawn in June 1964. The engine remained at Carlisle until September 1964, when it went to Connel's Scrapyard in Coatbridge, where it lingered until November 1964, before finally being cut up. The welded tender fitted when the engine was rebuilt is clearly seen, as is the AWS air tank (March 1959). As an original Patriot, the engine was unnamed until rebuilding and was named on 3 November 1948. The external condition of the engine is excellent, making it surprising that it was withdrawn. KEN TYLER

45532 Illustrious (rebuilt July 1948) is seen in the roundhouse at Carlisle Upperby on 29 February 1964. This was its home shed, having arrived in June 1962 from Saltley. It remained there until withdrawal in February 1964. The locomotive had been one of the various Class 7s that were allocated to Midland depots in 1959, when it was allocated to Nottingham. It was part of the mass transfer to Saltley in June 1961, then again part of the mass transfer to Carlisle Upperby in June 1962. The engine has both AWS (November 1959) and a speedometer drive (January 1961) and still has its nameplates, which was not common by 1964. The engine was withdrawn in February 1964 and was scrapped at Campbell's at Airdrie in January 1966, following storage at Carlisle Upperby. Ken Tyler

46118 Royal Welch Fusilier (rebuilt December 1946) is seen in store at Carlisle Upperby on 19 August 1964, by which time the engine had been withdrawn (in June 1964). The engine remained at Upperby until October 1964, when it was moved to Connell's Scrapyard at Calder. The engine parked behind is 45545 Planet, which had been withdrawn at the same time as 46118 and was to remain in store until September 1964, when it was scrapped. The rebuilt Scot displays its riveted tender and its AWS tank. Surprisingly, both engines were stored complete with nameplates and crests and both were in excellent external condition. The locomotive was one of twelve rebuilt Class 7s that had arrived en masse from Saltley in June 1962. KEN TYLER

very smartly presented. Upperby had a good reputation for turning out locos for football specials in excellent external condition. 46108 *Seaforth Highlander* was seen on the 12.28 Leeds–Manchester express on 16 June 1962. As the 1960s progressed, Upperby's engines could be seen anywhere, so on 24 August 1963, 46166 *London Rifle Brigade* was seen pulling the relief to the *'Pines Express'* from Wolverhampton northwards to Crewe.

Preston

Preston had a stud of Jubilees and a small number of rebuilt Class7s that worked Blackpool–Euston services, but the Jubilees had left by June 1961 and a couple of additional rebuilt Class 7s were moved into the depot. Preston shared some of the passenger diagrams to Blackpool with Camden, which included some Class 7 workings.

Some of the many Camden diagrams for the rebuilds included working some Blackpool services to Euston and return, Workington services from Preston and a heavy suburban train between Wigan and the Royal Ordnance Factory (ROF) at Chorley (Euxton) and return. This train was often loaded up to twelve coaches and ran for part of its journey on the West Coast Main Line, so needed Class 7 power. With the arrival of the English Electric Type 4s (Class 40) at Camden, a number of the Class 7 steam diagrams were changed and as there were now surplus Class 7s in the system a number were allocated to Preston to cover these former Camden diagrams.

However, a disastrous fire on 28 June 1960 caused the shed to close in September1961. One of the locos inside the shed when it caught fire was 46161 *King's Own*, which was photographed outside the shed with the paintwork badly scorched. The insurance claim allowed all of the locomotives damaged in the fire to be repaired at no cost to British Railways. The shed site continued to be used for some months after the fire, but the depot was finally closed on 12 September 1961 and as there was still a requirement for Class 7 power on the ROF trains the Class 7s were transferred to Springs Branch Wigan.

46168 The Girl Guide *(rebuilt April 1951) can be seen in store at Preston shed on 1 August 1964 and still has its front number plate. Other photographs taken at the time show that the locomotive still had its nameplates attached. The locomotive had been withdrawn from Springs Branch Wigan in May 1964 and after storage at Springs Branch between May and August it moved for a few weeks to Preston shed, before moving to Crewe Works, where it was scrapped at the end of August 1964. The engine had been at Springs Branch since September 1961 (and was the shed's longest rebuilt Class 7 resident), where its primary passenger use was on the heavy Royal Ordnance Factory trains to and from Chorley, as well as Manchester–Blackpool trains and Blackpool–Euston trains.* S.L. POGMORE

Rebuilt Class 7s at Preston

Loco	From	Date	To	Date	Comments
45735 *Comet*	Camden	September 1959	8A Edge Hill	September 1960	
46161 *King's Own*	Camden	September 1959	Bushbury	March 1961	
46161 *King's Own*	Bushbury	June 1961	Springs Branch	September 1961	
46165 *The Ranger (12th London Regiment)*	Upperby	November 1959	Springs Branch	September 1961	
46167 *The Hertfordshire Regiment*	Upperby	November 1959	Upperby	September 1960	
46167 *The Hertfordshire Regiment*	Bushbury	June 1961	Springs Branch	September 1961	
46168 *The Girl Guide*	Camden	September 1959	Springs Branch	September 1961	
46170 *British Legion*	Crewe North	December 1959	Crewe North	January 1960	

46161 King's Own (rebuilt October 1946) is seen at Preston shed following the major fire that destroyed the shed on 28 June 1960. The fire resulted in damage to a number of engines and the scorched paintwork and broken cab windows can be clearly seen. The engine was allocated to Preston (arriving in September 1959 from Camden) at the time and the shed's small allocation of Class 7 rebuilds covered a number of diagrams that had previously been covered by 1B Camden engines. Following the fire, the engine was sent to Crewe Works for inspection and repainting (the EHC logs a Light Classified repair between 29 June 1960 and 20 August 1960). The leading driving wheels are of the Stanier style, with the riveted balance weights.

The engine returned to 24K Preston before moving to 8F Springs Branch in August 1961, which took over from the now-closed Preston shed. A further move to Crewe North was made in June 1962, before a final move to 55A Leeds Holbeck in July 1962 and withdrawal in December 1962. Scrapping took place at Crewe Works in December 1963. The long gap between withdrawal and scrapping was due to the engine being stored at 55H Leeds Neville Hill from December 1962 until November 1963. STEPHENSON LOCO SOCIETY

Manchester Newton Heath

Newton Heath was coded 26A and was the former Lancashire and Yorkshire (L&Y) passenger depot on the north side of Manchester. It had passenger workings from Manchester Victoria to Glasgow, as well as working on the services from Liverpool to Leeds. The Saturday extras also included workings to Euston, Penrith (for Workington), as well as Manchester–Blackpool North and Manchester–Rochdale services. There were also Manchester–Leeds and Manchester–York newspaper trains.

For many years, these passenger workings had been covered by Jubilees, including the long-distance trains to Glasgow, but some Britannia 4-6-2s had been allocated to the depot for some of its express services in January 1960. With the transfer away of the Britannias in September 1961, some rebuilt Class 7s appeared that same month and they were primarily used on the Manchester –Glasgow trains, as well as on expresses to Blackpool. A couple of rebuilt Scots (46106 *Gordon Highlander* and 46137 *The Prince of Wales's Volunteers*

(South Lancashire) had previously arrived in 1960, but had been transferred away after nine months.

The Newton Heath crews, however, never really took to the rebuilt Scots and preferred the Jubilees on the Glasgow services, so the rebuilt Class 7s were more usually seen on the Manchester Victoria–Blackpool and Manchester Victoria–Southport expresses. The reign on the express passenger turns at the depot was short-lived, as in 1962 the depot had some English Electric Type 4s allocated and these took over the Glasgow trains, although on 18 August 1962 46133 *The Green Howards* was seen on a Manchester–Glasgow express. Other engines were seen further afield, with 46142 *The York and Lancaster Regiment* seen on the Lakes express on 1 August 1963.

On 13 and 14 January 1963, 46133 *The Green Howards* and 46140 *The King's Royal Rifle Corps* were borrowed by Bristol Barrow Road shed to cover an acute motive power shortage in the extreme winter of 1962–3, although this was a last gasp for 46133, as it was withdrawn a few weeks later.

45522 Prestatyn (rebuilt February 1949) and 46133 The Green Howards (rebuilt July 1944) are seen at Newton Heath in 1963, with The Green Howards already withdrawn (in February 1963), but remaining stored at the depot until April 1963, when it left for scrapping at Crewe Works in May 1963. The frontal view shows the differences between the fronts of a rebuilt Patriot and a rebuilt Scot. Prestatyn is fitted with the AWS 'bash plate' behind the front coupling. After ten years allocated to Leeds Holbeck, 46133 The Green Howards moved to the Midland Main Line, with stays at Trafford Park and Kentish Town, before arriving at Newton Heath in September 1961. 45522 Prestatyn arrived at Newton Heath in September 1961 and left for 9A Longsight in June 1963. Although nominally allocated to Longsight, the engine seemed to have been borrowed by Buxton and was used regularly on the fast Buxton–Manchester Central and return trains. It remained at Buxton after withdrawal in September 1964 and did not leave for scrap until May 1965.

Rebuilt Class 7s at Newton Heath

Loco	From	Date	To	Date	Comments
45522 *Prestatyn*	Kentish Town	September 1961	Longsight	June 1963	
46106 *Gordon Highlander*	Longsight	April 1960	Trafford Park	December 1960	
46133 *The Green Howards*	Kentish Town	September 1961	Withdrawn	February 1963	Stored at Newton Heath, February 1963 to April 1963 and then scrapped at Crewe Works, May 1963.
46137 *The Prince of Wales's Volunteers (South Lancashire)*	Longsight	April 1960	Trafford Park	December 1960	
46139 *The Welch Regiment*	Kentish Town	September 1961	Withdrawn	October 1962	Stored at Newton Heath, October 1962 to April 1963 and then scrapped at Crewe Works, May 1963.
46140 *The King's Royal Rifle Corps*	Kentish Town	September 1961	Longsight	June 1963	Spent sometime in store at Agecroft in the winter of 1962–3.
46142 *The York and Lancaster Regiment*	Kentish Town	September 1961	Longsight	June 1963	Spent sometime in store at Agecroft in the winter of 1962–3.

Rebuilt Class 7s on the Midland Main Line

The Midland Main Line had been the preserve of the ex-LMS Jubilee 4-6-0, but double-heading had not been eliminated and there had also been increasing complaints regarding timekeeping. The Midland Main Line timetable was revised in June 1957 and to eliminate double-heading six rebuilt Class 7s were transferred on loan to Kentish Town (14B) in October 1957, along with three Britannias. In July 1958, the Scots returned to the Western Division of the London Midland Region and additional Britannias were allocated to Trafford Park.

In September 1959, six rebuilt Scots arrived at Kentish Town (three from Edge Hill and three from Longsight), followed in November 1959 by two rebuilt Scots and a rebuilt Patriot from Camden. At the same time, three rebuilt Scots were allocated to Nottingham (from Camden and Edge Hill). In February 1960, Sheffield Millhouses received five rebuilt Scots and two rebuilt Patriots. In the winter of 1960–1, seven rebuilt Scots and one rebuilt Patriot were allocated to Trafford Park, replacing the Britannias at the depot.

However, the reign of the rebuilt Class 7s was short-lived, as from October 1960 the Sulzer Type 4s (Class 45 and 46) were being introduced and by the end of March 1961 thirty-six of these Type 4s were in service on the Midland Main Line. By the summer of 1961, the Class 45s had replaced the rebuilt Class 7s on the Leeds–Birmingham–Bristol services. This resulted in the transfer of seven rebuilt Scots and two rebuilt Patriots to Saltley, where they worked many of the many extra Saturday trains, but during the week they could be regularly seen on freight services.

London Kentish Town

Kentish Town, coded 14B, received an allocation of rebuilt Class 7s in October 1957 for the principal express passenger workings from St Pancras to Nottingham, Derby and Sheffield. Prior to the allocation, the depot had relied on its Jubilee 4-6-0s, but with

46163 Civil Service Rifleman (rebuilt October 1953) is seen at Miller's Dale in the Derbyshire Peak District pulling a Manchester–St Pancras train. A fleet of rebuilt Class 7s had arrived on the Midland region in 1959. Strangely, 46163 was at the time not a Midland depot loco, as during 1960 it was based at Crewe, Holyhead, Preston and Llandudno Junction. BLENCOWE COLLECTION

46142 The York and Lancaster Regiment (rebuilt February 1951) is seen at Blackwell Mill in the Peak District with the 2.00pm from Manchester Exchange on 20 September 1959. 46142 had arrived at 14B Kentish Town in the same month as the photograph. The engine is unusual for a rebuilt Scot, being fitted with a welded tender and the rolling stock in the train is a real mixture of BR Mk1 and ex-LNER stock. E.R. MORTEN

Rebuilt Class 7s at Kentish Town

Loco	From	Date	To	Date	Comments
45522 *Prestatyn*	Camden	October 1959	Newton Heath	September 1961	
46110 *Grenadier Guardsman*	Crewe North	October 1957	Camden	July 1958	On loan.
46116 *Irish Guardsman*	Carlisle Upperby	October 1957	Camden	July 1958	On loan.
46123 *Royal Irish Fusilier*		September 1959	Saltley	June 1961	
46127 *Old Contemptibles*	Crewe North	October 1957	Holyhead	July 1958	On loan.
46131 *The Royal Warwickshire Regiment*	Longsight	October 1957	Longsight	November 1958	On loan.
46132 *The King's Regiment Liverpool*	Longsight	September 1959	Saltley	October 1961	
46139 *The Welch Regiment*	Camden	November 1959	Newton Heath	September 1961	
46140 *The King's Royal Rifle Corps*	Edge Hill	September 1959	Nottingham	November 1959	
46140 *The King's Royal Rifle Corps*	Nottingham	January 1960	Newton Heath	September 1961	
46142 *The York and Lancaster Regiment*	Edge Hill	September 1959	Newton Heath	September 1961	
46152 *The King's Dragoon Guardsman*	Edge Hill	October 1957	Camden	July 1958	On loan.
46157 *The Royal Artilleryman*	Edge Hill	October 1957	Crewe North	June 1958	On loan.
	Edge Hill	August 1959	Newton Heath	September 1961	
46160 *Queen Victoria's Rifleman*	Longsight	September 1959	Saltley	June 1961	
46162 *Queen's Westminster Rifleman*	Camden	November 1959	Saltley	June 1961	Initially on loan.

The pioneer rebuilt Scot 46103 Royal Scots Fusilier (rebuilt June 1943) is seen at the Derby shed after its transfer to 14B Kentish Town for work on the Midland Main Line in October 1958, where it remained until it moved to Saltley in June 1961. The loco had been a long-term resident at 55A Leeds Holbeck, having arrived shortly after its rebuilding in August 1943 and staying at the shed until its move to Kentish Town. After the move to Saltley, it went to Carlisle Upperby in June 1962 and back to Leeds Holbeck in September 1962, before withdrawal in December 1962. The locomotive was never fitted with AWS, but did acquire a speedometer drive in the late 1960s. COLOUR-RAIL

the introduction of a new timetable and the availability of rebuilt Class 7s from the Western Division it was possible to use these engines. Initially, the engines were on loan, but a substantial allocation was made in 1959 and they lasted until October 1961 when the last was reallocated elsewhere. The depot would continue to receive rebuilt Class 7s, as visiting engines from Nottingham, Trafford Park and Millhouses (and later Canklow and Darnell) could be seen. The Sulzer Type 4s encountered lots of problems and it was not unusual to see steam at the head of services arriving at St Pancras.

Sheffield Millhouses

Sheffield Millhouses was coded 25A and was part of London Midland Region until it was transferred to North Eastern Region in February 1958 as 41C and closed in January 1962. It was the Midland/LMS depot for Sheffield Midland and had responsibility for Sheffield–St Pancras services, as well as to Birmingham and Bristol. Prior to the arrival of the rebuilt Class7s, the depot had possessed a stud of Jubilees.

The depot closed in 1 January 1962 and the locomotives were moved to Sheffield Canklow.

Rebuilt Class 7s at Sheffield Millhouses

Loco	From	Date	To	Date	Comments
45514 Holyhead	Camden	February 1960	Withdrawal	May 1961	The first rebuild to be withdrawn. It was alleged that the loco had entered Crewe Works for overhaul, but when stripped down the loco was found to have badly cracked frames and was withdrawn instead.
45536 Private W. Wood, V.C.	Edge Hill	February 1960	Canklow	November 1961	
46131 The Royal Warwickshire Regiment	Edge Hill	February 1960	Canklow	November 1961	
46147 The Northamptonshire Regiment	Crewe	February 1960	Canklow	November 1961	
46148 The Manchester Regiment	Upperby	February 1960	Canklow	November 1961	
46151 The Royal Horse Guardsman	Longsight	February 1960	Canklow	November 1961	
46164 The Artists' Rifleman	Crewe North	February 1960	Canklow	November 1961	

Canklow

With the closure of Millhouses 41C, the rebuilt Class 7s were transferred to Canklow 41D, although by this time the Sulzer Type 4s had been introduced, reducing the amount of passenger work for the engines.

Rebuilt Class 7s at Canklow

Loco	From	Date	To	Date	Comments
45536 Private W. Wood, V.C.	Millhouses	November 1961	Darnel	June 1962	
46131 The Royal Warwickshire Regiment	Millhouses	November 1961	Llandudno Junction	February 1962	
46147 The Northamptonshire Regiment	Millhouses	November 1961	Willesden	February 1962	
46148 The Manchester Regiment	Millhouses	November 1961	Crewe North	April 1962	
46151 The Royal Horse Guardsman	Millhouses	November 1961	Darnel	June 1962	Seen stored at Staveley shed April 1962.
46164 The Artists' Rifleman	Millhouses	November 1961	Darnel (41A)	June 1962	

Sheffield Darnall

Darnall took over the ex-Millhouses rebuilt Class 7s, but their time at the depot was short-lived, as the majority of the Midland Main Line services from Sheffield had been dieselized in the summer of 1961 and the only passenger work would be to substitute for diesel non-availability . All the Darnall examples were withdrawn by the end of 1962, just in time for the exceptionally bad winter into 1963.

Darnall closed on 17 June 1963, but by this date all the rebuilt Class 7s had already been withdrawn.

45536 Private W. Wood VC (rebuilt November 1948) is seen in store at Sheffield Darnall shed after withdrawal in October 1962. It remained there until it moved to further storage at Staveley in February 1964 and was scrapped at Crewe Works in March 1964. The engine had moved to Millhouses in February 1960 for working the Midland Main Line services from Sheffield and with the closure of Millhouses moved to Canklow, before moving to Darnall in June 1962. The engine has a riveted tender, which it acquired in July 1960 and has AWS (February 1959) and a speedometer drive (February 1961). The withdrawal came as a surprise to the Darnell boiler smiths, as on the day the engine was withdrawn they had just completed some boiler repairs when instructions came to withdraw the engine. G. SHARPE

Rebuilt Class 7s at Darnall

Loco	From	Date	To	Date	Comments
45536 *Private W. Wood, V.C.*	Canklow	June 1962	Withdrawn	October 1962	Boiler work completed the same day that the engine was withdrawn.
46151 *The Royal Horse Guardsman*	Canklow	June 1962	Withdrawn	December 1962	Stored at Darnall, December 1962 to July 1963. Scrapped at Crewe Works August 1963.
46164 *The Artists' Rifleman*	Canklow	June 1962	Withdrawn	December 1962	Stored Crewe Works, January 1963 to February 1963 and scrapped there March 1963.

Nottingham

Nottingham shed was coded 16A. The locos were primarily used on the *'Robin Hood'*, a Nottingham–St Pancras service leaving Nottingham at 8.00am and returning from St Pancras in the evening, a round trip of 250 miles (400km) per day. The depot's engines could also be seen on St Pancras–Leicester workings.

As with many of the rebuilt 7s allocated to the Midland Main Line, the reign of the engines at Nottingham did not last long, as the introduction of the Sulzer Type 4s from early 1961 meant that the steam locomotives soon drifted away to Saltley and onwards to oblivion via Annesley and Carlisle Upperby. With the advent of the Sulzer Type 4s, Nottingham's rebuilt Class 7s were seen on various trains. For example, on 7 May 1961 46112 *Sherwood Forester* pulled an Accrington–Wembley football excursion and on 10 February 1962 a Leeds–Edinburgh express. 46100 *Royal Scot* was a lucky survivor, as it was sold in October 1962 to Butlin's for future display at the Skegness holiday camp.

46100 Royal Scot (rebuilt June 1950) is seen at Wellingborough on the Midland Main Line. The engine had moved to Nottingham in November 1959 and remained there (except for a move to Derby between June 1961 and August 1962), before being withdrawn in October 1962, although it was noted in store at Nottingham in September 1962. The engine has AWS (May 1959) and a speedometer drive (October 1960) and has a riveted tender that it acquired in 1936. The name appears to have the red background that was noted on the engine when at Nottingham. The engine does not have the overhead warning flashes, which it would have acquired at sometime in 1960. The engine was bought by Butlin's and moved to Crewe Works in October 1962, where it was overhauled and received its LMS crimson livery before leaving the Works in May 1963.

45532 Illustrious (rebuilt July 1948) is seen at Nottingham shed. It was one of a number of rebuilt Class 7s that were allocated to Midland region depots from 1959, with engines going to Kentish Town, Nottingham, Sheffield Millhouses and Trafford Park. 45532 Illustrious arrived at Nottingham in November 1959 and then moved to Saltley in June 1961, before a move to Derby in April 1962 and a final move to Carlisle Upperby in June 1962. The engine was withdrawn in February 1964, before being scrapped at G.H. Campbell, Airdrie, in December 1964. The driving wheels are the early Stanier type with the webbing on the spokes around the crankpins and the bogie wheels are Stanier. AWS was fitted in November 1959 and a speedometer in January 1961. The welded tender is the one fitted when the engine was rebuilt in 1948. PETER GROOM

46111 Royal Fusilier (rebuilt October 1947) is seen inside the roundhouse at 16A Nottingham shed on 6 June 1963, when the engine was officially allocated to Annesley depot. The engine has lost its nameplate and has AWS and a speedometer, as well as the small badge holder for the driver's name. Prior to moving to Annesley (in February 1963), 46111 had been resident at Willesden, arriving in June 1961. Nottingham had acquired a small stud of Class 7s in November 1959, but all had been transferred away or withdrawn by August 1962. The engine was borrowed for main line passenger working during the extreme winter of 1962–3 to cover for diesel failures. COLTAS TRUST

Rebuilt Class 7s at Nottingham

Loco	From	Date	To	Date	Comments
45532 *Illustrious*	Camden	December 1959	Upperby	July 1961	Seen with red-backed nameplate in February 1960.
46100 *Royal Scot*	Camden	December 1959	Derby	June 1961	
46100 *Royal Scot*	Derby	August 1962	Withdrawn	October 1962	To Crewe Works, October 1962 to May 1963 for restoration.
46112 *Sherwood Forester*	Leeds Holbeck	Late 1959	Trafford Park	December 1960	
46118 *Royal Welch Fusilier*	Camden	Late 1959	Saltley	August 1961	
46157 *The Royal Artilleryman*	Edge Hill	Late 1959	Saltley	June 1961	

Derby

Derby was coded 17A and was part of London Midland Region until it closed in March 1967. The depot was more usually associated with an allocation of Jubilees, with eleven on its books in early 1961. The principal diagrams for the Jubilees were to Bristol, Sheffield, St Pancras, Leeds and Bath. However, in the summer of 1961 many of these workings were taken over by Sulzer Type 4s (initially mainly the Manchester Central–St Pancras trains), so the author has not been able to identify which specific workings the rebuilt Class 7s were responsible for, although it may have been to act as standby for any problems with the newly introduced Sulzer Type 4s. 46106 *Gordon Highlander* was seen at Wellingborough shed on 20 August 1961 and is presumed to have worked a freight to the depot.

46100 Royal Scot (rebuilt June 1950) is seen at the Derby shed when the engine was based on the Midland division and was seen on Midland Main Line workings. The engine arrived at Nottingham in November 1959 from Camden and stayed until June 1961, when it moved to Derby, before returning to Nottingham in August 1962. The engine was withdrawn from Nottingham in October 1962 for preservation and went to Crewe Works for a cosmetic restoration into an LMS livery of Crimson Lake. The engine has acquired the usual modifications (AWS in May 1959 and speedometer drive in October 1960). The engine has also been fitted with a badge holder for the driver's name on the cab side, but these were never popular and do not seem to have been used.

Rebuilt Class 7s at Derby

Loco	From	Date	To	Date	Comments
45514 *Holyhead*	Millhouses	May 1961	Withdrawn	May 1961	Never arrived as withdrawn at Crewe Works after stripping down.
46100 *Royal Scot*	Nottingham	June 1961	Nottingham	August 1962	
46106 *Gordon Highlander*	Trafford Park	June 1961	Saltley	September 1961	
46118 *Royal Welch Fusilier*	Nottingham	June 1961	Saltley	August 1961	

Most of the depot's Jubilees were moved away in early 1961, with some remaining until November 1961, although some returned in January 1963 during the appalling weather in the winter of 1962–3. Two of the rebuilt Class 7s joined the mass migration of the Midland division's Class 7s to Saltley in the autumn of 1961.

Manchester Trafford Park

Trafford Park was coded 13A London Midland Region in 1950 and renumbered 9E (as a sub-shed of 9A Longsight) in May 1950. It closed on 4 March 1968. Trafford Park was not only responsible for providing Class 7 power for express trains from Manchester Central to London St Pancras, but also for the

46158 *The Loyal Regiment* (rebuilt September 1952) sits inside its home depot of 9E Trafford Park on 24 April 1961, having arrived at the depot in December 1960 when a number of engines were moved to work the Midland Main Line services from Manchester to St Pancras. as well as the Leeds and Sheffield to St Pancras services. 46158 was considered to be the best of the rebuilt Scots by the crews who fired and drove them. The loco has AWS (January 1959) and a speedometer drive (July 1960) and also has the small bracket for the driver's name on the cab side. *COLTAS TRUST*

46142 *The York and Lancaster Regiment (rebuilt February 1951) is seen at Trafford Park shed and displays a red-painted backing to the nameplate, although the crest above the name is missing. The engine at the time was allocated to Longsight, but Trafford Park regularly borrowed rebuilt Class 7s from Longsight and would use them on the passenger services to and from Buxton and Manchester Central. The engine would stand spare at Trafford Park in case of diesel failure.* G. SHARPE

Manchester Central–Buxton express in the evening and one from Buxton– Manchester Central in the morning.

With the replacement of the rebuilt Class 7 by the Sulzer Type 4s in the summer of 1961, many of the rebuilds had little regular work, but it was usual to have one of them standing 'spare' at the shed to cover for any diesel failure. The Buxton workings in the morning and evening remained a rebuilt Class 7 working, with the engine standing spare at Trafford Park for the day. When Longsight depot became responsible for the Buxton workings (again with rebuilt Class 7s), the practice of the engine standing spare a Trafford Park continued.

Rebuilt Class 7s at Trafford Park

Loco	From	Date	To	Date	Comments
46106 *Gordon Highlander*	Crewe North	December 1960	Derby	June 1961	
46112 *Sherwood Forester*	Nottingham	December 1960	Annesley	February 1962	
46122 *Royal Ulster Rifleman*	Willesden	January 1960	Saltley	June 1961	
46137 *The Prince of Wales's Volunteers (South Lancashire)*	Newton Heath	January 1961	Saltley	August 1961	Initially on loan.
46141 *The North Staffordshire Regiment*	Willesden	December 1960	Saltley	June 1961	
46143 *The South Staffordshire Regiment*	Bushbury	December 1960	Annesley	February 1962	
46153 *The Royal Dragoon*	Bushbury	January 1960	Annesley	September 1962	Initially on loan.
46158 *The Loyal Regiment*	Willesden	December 1960	Annesley	February 1962	

CHAPTER SIX

REBUILT CLASS 7s TO THE RESCUE – THE SEVERE WINTER OF 1962–3

The winter of 1962–3 was one of the worst of the twentieth century, with sub-zero temperatures from mid-December 1962 through to April 1963. It caused havoc on the railways, particularly to the new diesel fleets. The bad weather started in November, with blizzards sweeping across parts of the country and in early December freezing fog blanketed parts of the country. At the end of December 1962, a blizzard struck the south of England, piling snow up to 20ft (6m) deep. The freezing temperature meant that the snow lasted for up to two months in some areas and January 1963 was the coldest month of the twentieth century, with the lowest temperature being –19°C and the sea froze over in Kent.

In February 1963, more snow arrived and a thirty-six hour blizzard caused heavy drifting. Freezing fog on some days also caused disruption. The thaw did not start until early March and 6 March was the first day of the year without a frost. Reliability of the new diesels had been an issue before the extreme weather, but the sub-zero temperatures also played havoc with steam-heating boilers and also led to fuel freezing up and engines shutting down.

The result was that many of the rebuilt Class 7s that had been banished to lesser workings suddenly found themselves back on express passenger services. On the West Coast Main Line, on 18 November 1962 46130 *The West Yorkshire Regiment* (55A Leeds Holbeck) was seen on the sleeper train to Glasgow and on the Midland Main Line 46140 *The King's Royal Rifle Corps* (26A Newton Heath) was seen on a St Pancras–Leicester express on 28 December. On 4 November, 46144 *Honourable Artillery Company of Holyhead* was seen on the 9.42 Newcastle–Liverpool

from Leeds and on 10 January 1963, 46126 *Royal Army Service Corps* was seen on a Gloucester–Sheffield train at Birmingham.

Diesel availability was poor, with both the English Electric Type 4s and the Sulzer Type 4s having less than 50 per cent availability, putting pressure on depots to use steam as replacements. On 17 January 1963, 45532 *Illustrious* (12B) was seen at Preston on the up Caledonian to London, running two hours late (having replaced a Class 40 at Carlisle that had a defective train-heating boiler) and stopping for water, as the water scoop on the tender was out of order.

The Liverpool–Newcastle service, which in steam days would have been worked as far as Leeds by a rebuilt Class 7, found itself reverting to steam, so on 10 and 11 December 1962 46114 *Coldstream Guardsman* (6J) and 46168 *The Girl Guide* (8F) worked between Liverpool and Newcastle. In January 1963, the service was pulled by rebuilt Class 7s on many occasions, with 46150 *The Life Guardsman* (6J) on 6 January, 46155 *The Lancer* (6G) on 10 January, 45545 *Planet* (12B) on 21 January, 45522 *Prestatyn* (26A) on 22 and 23 January, 46149 *The Middlesex Regiment* (9A) on 25 January and 46110 *Grenadier Guardsman* (8A) on 26 January.

Matters only marginally improved in February, when on 4 February 45512 *Bunsen* (12B) took the 11.00am Liverpool–Newcastle as far as Leeds and on 16 February the Liverpool–Newcastle was worked all the way through to Newcastle by 45531 *Sir Frederick Harrison* (8A), only losing sixteen minutes on the booked timings and returning back to Liverpool the following day on a freight.

*46150 The Life Guardsman
(rebuilt December 1945) is seen at
Bristol Barrow Road shed in 1961
and a long way from its home depot
of Holyhead. Rebuilt Class 7s were
regular visitors to Bristol, with
services from Leeds, Sheffield and
Birmingham. During the summer of
1961, the rebuilt Class 7s were
replaced by Sulzer Type 4s (Class
45), although they had to substitute
regularly for diesel non-availability.
AWS was fitted in May 1959 and a
speedometer in June 1961. The
severe winter of 1962–3 saw
Barrow Road borrow rebuilt Class 7s
from across the country.*

The situation was still difficult in March and on 13 March 46110 *Grenadier Guardsman* (8A) was seen on the 11.00am Liverpool–Newcastle and on 29 March 46116 *Irish Guardsman* (12A) was seen on the Manchester–Newcastle service.

Availability of the diesels fell from 80 per cent to less than 50 per cent in January 1963 and Bristol Barrow Road had to borrow rebuilt Class 7s to cover diesel diagrams up to Birmingham, Derby and Sheffield. Locos seen included 46133 *The Green Howards* (26A Newton Heath), 46140 *The King's Royal Rifle Corps* (26A Newton Heath) and 46149 *The Middlesex Regiment* (9A Longsight). 46128 *The Lovat Scouts* (12A Kingmoor) was being used as diesel standby on 1 January. The fact that all of these borrowed engines came from sheds many hundreds of miles away showed how serious the situation was and how desperate shed masters were getting to keep the trains moving.

In January, 46149 *The Middlesex Regiment* was seen on the up Thames Clyde forwards from Sheffield, and Carlisle Upperby took two withdrawn locomotives from the scrap line (46106 *Gordon Highlander* and 46127 *Old Contemptibles*) for service for a few weeks, so hard pressed were the operating authorities. 46106 reached Cricklewood in London and was also seen on

a parcels train at Preston and 46127 was seen on the North Wales Coast on a freight.

Polmadie, which had withdrawn its own fleet of rebuilt Scots at the end of 1962, was forced in February to borrow 46166 *London Rifle Brigade* from Crewe North for a few days to cover some of its express passenger workings. Other services seen in February were the 3.20 Nottingham–St Pancras with 46111 *Royal Fusilier* (Annesley) and on 4 February 46142 *The York and Lancaster Regiment* (26A Newton Heath) was put on a Glasgow–Manchester service at Preston and on 16 February 1963 46160 *Queen Victoria's Rifleman* (12B Upperby) was seen on the 5.5pm Blackpool–Euston.

In March, many of the Wolverhampton–Euston trains reverted to steam and in April 46101 *Royal Scots Grey*, 46111 *Royal Fusilier* and 46158 *The Loyal Regiment* (all 16B Annesley locomotives) were observed at Nottingham Midland station.

By April, the severe weather was past, but 46142 *The York and Lancaster Regiment* (26A Newton Heath) was seen on the 9.43 Liverpool Exchange–Glasgow express and worked the train throughout. It was in poor external condition, but in good mechanical order.

CHAPTER SEVEN

DECLINE AND WITHDRAWAL

In 1958, British Railways commenced the introduction of larger diesel locomotives on to the routes operated by the rebuilt Class 7s. The English Electric Type 4 (Class 40s) were introduced as follows:

- 1958 – ten
- 1959 – forty
- 1960 – seventy
- 1961 – ten
- 1962 – seventy-five.

This gave a total of 205 allocated to the LM, ER and Scottish Regions. More important was the allocation on the West Coast Main Line to the following depots in 1961:

- Camden (London) – twenty-three
- Longsight (Manchester) – fourteen
- Edge Hill (Liverpool) – sixteen
- Carlisle (Upperby) – eight
- Newton Heath (Manchester) – one
- Crewe North – twenty-seven.

All of these depots had extensive allocations of rebuilt Class 7s and many of their duties were either taken over by diesels, or by Duchess Pacifics cascaded down from the top-class trains by the diesels.

By 1961, on the West Coast Main Line the English Electric Type 4 (later Class 40) had replaced the Duchess Pacifics and rebuilt Class 7s on the top workings and of course a smaller number of diesels could

45545 Planet *(rebuilt November 1948) is seen at platform 2 at Manchester Victoria on 4 May 1963 with the 10.50am local train to Wigan. The engine shows a Carlisle Upperby plate and is a good example of a 'fill-in' turn, so if an engine had worked into Manchester in the morning and was not booked to return until late in the afternoon, it could be used on more local services and return back to the depot ready for its booked return service. At the time, there were regular fast freight workings from Carlisle to Manchester (Liverpool Road) and return that were regularly worked by rebuilt Class 7s. The engine is fitted with AWS (March 1959) and speedometer drive (October 1960) and had acquired its smoke deflectors in February 1952. It has its welded tender that it acquired at rebuilding and retained until its withdrawal in June 1964 from 12B Carlisle Upperby. JIM CARTER*

replace a given number of steam locomotives. The expresses between Euston–Liverpool and Manchester–Glasgow were soon principally in the hands of the Class 40s and the rebuilt Class 7s were no longer regularly diagrammed for the top-link trains. The new diagrams for the diesels were more complex than previously used for steam locomotives, so as an example on day one of a cycle, one of 8A Edge Hill's Class 40s would work a Liverpool–Euston and would then work a Euston–Manchester express, staying overnight at 9A Longsight. The following day it would work a Manchester–Euston train. It would then work to Birmingham and Manchester as far as Crewe, where it would stay overnight. The following morning it would take over a Manchester–London express at Crewe and then return on a Euston–Liverpool express. The loco would then repeat the three-day cycle.

In addition, the electrification between Manchester and Crewe was complete, so many services were electric-hauled as far as Crewe, where the train would hand over to either a diesel or a steam locomotive. However, the initial unreliability of the new diesels

and the severe winter of 1962–3 gave ample opportunity for the rebuilt Class 7s still to be seen on express trains.

The other main class of Type 4 diesels was the group collectively known as 'Peaks' after the first ten named examples. They were manufactured and introduced in a similar time frame to the Class 40s:

- 1959 – ten
- 1960 – fifteen
- 1962 – ninety-five
- 1962 – seventy
- 1963 – one hundred and forty-five.

On the ex-Midland Main Line and the route from Leeds down to Bristol (via Sheffield and Birmingham), the Peaks were introduced for the summer service in 1961 and examples were allocated to Leeds Holbeck, Derby and Gateshead. These replaced steam locos on the workings from Leeds to Sheffield, Derby, Birmingham and Bristol and also on the Newcastle–Liverpool services, which had been in the

46115 Scots Guardsman *(rebuilt August 1947). Scots Guardsman is seen leaving Helliefield on an RCTS special on 13 February 1965. It had been specially cleaned and replica nameplates fitted at Crewe North the week before the tour. The engine originally selected had been* 46160 *Queen Victoria's Rifleman, but it had run a hot bearing the week before, so 46115 was selected instead. The engine survived until the end of the year, before going into preservation.* G. SHARPE

hands of Edge Hill Class 7 rebuilds as far as Leeds. The Midland Main Line services of Sheffield–St Pancras via Derby, Manchester–St Pancras and Nottingham–St Pancras were also Peak-hauled, impacting on the allocations of Class 7s at Kentish Town, Sheffield (Darnell), Nottingham and Trafford Park. Once the reliability of the new diesels had been established, the Class7 rebuilds suddenly had no work, so the decision was made to reallocate them.

Reallocations

Leicester (GCR)

This ex-GCR shed had allocations of rebuilt Class 7s on two occasions. In the spring of 1961, Leicester City Football Club played in the FA Cup Final at Wembley, so a large number of special trains were run on the GCR route and for these heavy specials a number of rebuilt Class 7s were loaned to Leicester GCR shed. These were 45532 *Illustrious* (16A), 45540 *Sir Robert Turnball* (14B), 46149 *The King's Royal Rifle Corps* and 46160 *Queen Victoria's Rifleman* (14B). The engines arrived at Leicester a few days before and were cleaned up for the Saturday. The engines were on shed on the Sunday before being returned to their 'home' depots. 45540 was held as a standby engine and was not used on any of the football specials, so later worked on the Saturday on the York–Bournemouth as far as Banbury.

In May 1962, two rebuilt Scots were allocated to Leicester for working the heavy Newcastle–Bournemouth train between Leicester and Banbury. However, the use of the Class 7s on this train was short-lived, as it was decided that the English Electric Type 3, which brought the train into Leicester from Sheffield,

would continue to Banbury and return to Sheffield on the balancing working. The Scots were then transferred away after a stay of only a few weeks. They stayed long enough to have 15C shed plates placed on the smoke-box door and a number of photographs show 46118 with its 15C shed plate. The Leicester men were of the opinion that if the rebuilt Class 7s that were sent to Annesley had instead been allocated to Leicester, they would have been kept in better condition.

Wigan Springs Branch

With the closure of Preston, the Class 7 locos were reallocated to Springs Branch to work Preston's remaining Class 7 diagrams, such as the Workington–Euston (from Preston), and the locos were seen on other passenger trains such as Manchester Piccadilly–Euston and Southport–Manchester expresses. They were also used on the heavy ROF trains to and from Wigan. Being allocated to the North West, the locos were also used on summer specials to Blackpool. As many of the workings from Springs Branch were freight, it was not unusual for the rebuilt 7s to be seen on freight trains, so 45521 *Rhyl* was observed on the Springs Branch Fleetwood Wyre Dock freight (mainly coal) in June 1963 and 46115 *Scots Guardsman* was seen on the Fleetwood Wyre Dock–Broad Street (London) fish in June 1964.

The locos would be borrowed by depots wanting Class 7 power to cover for diesel non-availability and as an example 46168 *The Girl Guide* (Springs Branch) was appropriated on 11 November 1963 for the 9.00am Liverpool–Newcastle service, where it only dropped three minutes on the diesel timings as far as Huddersfield and 46165 *The Ranger* was seen on the Manchester Piccadilly–Euston express on 1 July 1962. The ROF trains were discontinued at the end of August 1964.

1962 Allocations of Rebuilt Class 7s to Leicester

Loco	From	Date	To	Date
46106 *Gordon Highlander*	21A Saltley	May 1962	21A Saltley	May 1962
46118 *Irish Guardsman*	21A Saltley	May 1962	21A Saltley	May 1962

Allocations of Rebuilt Class 7s to Springs Branch

Loco	From	Date	To	Date	Comments
45521 *Rhyl*	Edge Hill	16 September 1961	Withdrawn	September 1963	Stored at Springs Branch, October 1963 and scrapped at Crewe Works November 1963.
45531 *Sir Frederick Harrison*	Edge Hill	October 1963	Carlisle Upperby	July 1964	
46110 *Grenadier Guardsman*	Edge Hill	July 1963	Kingmoor	July 1963	
46115 *Scots Guardsman*	Longsight	June 1964	Carlisle Upperby	July 1964	Only allocated for seven weeks.
46128 *The Lovat Scouts*	Crewe North		Carlisle Kingmoor	May 1962	Only stayed a few weeks before allocated to 12A.
46161 *King's Own*	Preston	September 1961	Crewe North	July 1962	
46165 *The Ranger (12th London Regiment)*	Preston	September 1961	Crewe North	August 1962	
46167 *The Hertfordshire Regiment*	Preston	September 1961	Crewe North	June 1962	
46168 *The Girl Guide*	Preston	September 1961	Withdrawn	May 1964	Stored on shed May 1964 to August 1964, then stored Preston shed August 1964, before scrapping at Crewe Works August 1964.

45521 Rhyl (rebuilt October 1946) sits at Carlisle station on 12 August 1961 waiting to take over a train to the south. Carlisle was a changeover point for many expresses, both north- and southbound. The engine was allocated to 8A Edge Hill, but was soon to move in September 1961 to Springs Branch Wigan, where it remained until withdrawal in September 1963. It was scrapped at Crewe Works in November 1963. It is fitted with AWS (December 1959) and the engine carries a welded tender that it retained until withdrawal.

N. PREEDY

45531 Sir Frederick Harrison *(rebuilt December 1947) is seen on 27 July 1964 on freight when the engine was allocated to Carlisle Kingmoor, where it had arrived in September 1963 from Springs Branch Wigan. It remained at Carlisle, swapping between Upperby and Kingmoor, and was withdrawn from Kingmoor in November 1965. It remained in storage at Kingmoor until a move to Campbell's Scrapyard, Airdrie, in January 1966, where it lingered until March 1966. The engine had received an overhaul at Crewe Works in 1963 and is seen fitted with a riveted tender that it acquired in 1956 and has the top smokebox lamp bracket relocated. It still carries its nameplate.* ALAN HEYES

46168 The Girl Guide *(rebuilt 1946) is seen at Springs Branch Wigan on 4 August 1962. It is pulling the empty stock from the Royal Ordinance Factory (Chorley) passenger train, which it is taking to North sidings ready for the service the following day. The heavy (up to twelve coaches) ROF and the fast timing was one of the reasons for allocating rebuilt Class 7s to Wigan. The engine was allocated to Springs Branch Wigan in September 1961 (from Preston), but the shed plate shown here is 8F for Wigan, so either the date for the photo is wrong, or the engine arrived at Wigan a month earlier than officially recorded. At the time, the shed at Preston was only a shell following the fire in 1960, so perhaps the engines moved before Preston officially closed in September 1961. The front smokebox number plate is also missing (and the engine has been given its original LMS number); however, photos taken after this date show the number plate reinstated.* COLTAS TRUST

Bradford Low Moor

Bradford Low Moor was in the North East Region, with the shed code 56F. It was another depot that had not previously been allocated rebuilt Class 7s, but it received the entire allocation from Leeds Holbeck in September 1961 and most of the locomotives arrived in excellent condition. The close proximity to Leeds also meant that in the event of any major problems with the Peaks, Holbeck could borrow the locos back from Low Moor.

46109 Royal Engineer *(rebuilt July 1943) sits at its new home shed of 56F Low Moor (Bradford). The engine had been allocated to 55A Leeds Holbeck from August 1948 following rebuilding and remained at the shed until September 1961, when it was transferred to Low Moor until June 1962. It then returned to Holbeck, remaining there until withdrawal in December 1962. The rebuilt Scots at Low Moor where used on a heavy stopping train to Southport and back and also for specials (as indicated by the reporting number 1X 42 on the front).*
Following withdrawal, the engine was stored at 55H Leeds Neville Hill from January 1963 until September 1963, when it was towed in convoy with 46130 The West Yorkshire Regiment *and pulled by condemned 46145* The Duke of Wellington's Regiment (West Riding)*, especially steamed to Crewe Works in September 1963.* G. SHARPE

Allocations of Rebuilt Class 7s to Low Moor

Loco	From	Date	To	Date
46109 *Royal Engineer*	Holbeck	Week ending 16/9/61	Holbeck	June 1962
46113 *Cameronian*	Holbeck	Week ending 16/9/61	Mirfield	January 1962
46117 *Welsh Guardsman*	Holbeck	Week ending 16/9/61	Mirfield	January 1962
46130 *The West Yorkshire Regiment*	Holbeck	Week ending 16/9/61	Holbeck	June 1962
46145 *The Duke of Wellington's Regiment (West Riding)*	Holbeck	Week ending 16/9/61	Mirfield	January 1962

Previously, the depot had only Stanier Class 5s for its principal workings; however, the rebuilt Class7s were not able to use all the routes that the Class 5s could use, so the work available was restricted.

The Low Moor crews noted that all of these locos were in excellent condition, with the exception of 46113 *Cameronian*, which was considered a poor steamer.

The principal passenger working at the depot was a Monday to Friday 7.05am Bradford–Southport passenger train (usually loaded to ten coaches), which stopped at forty stations on the outward-leg stations between Bradford and Southport. The return duty was a 1.10pm Southport–Bradford express (also a ten-coach train) calling at Wigan, Manchester, Rochdale and Halifax, a diagram that certainly taxed both the locomotive and crews. When the rebuilt Scots were allocated to the depot they were always given the Southport diagram and the engines impressed the Low Moor crew. One of the Low Moor fireman, Granville Dobson, recollected that 46145 *The Duke of Wellington's Regiment* (West Riding) was regularly diagrammed for this service and he swapped duties with other Low Moor firemen to get as many trips on the locomotive as possible. 46145 *The Duke of Wellington's Regiment* (West Riding) was in excellent condition and according to Granville had a superb ride – the memory of some trips with this engine on the Southport diagram still live fresh in his memory.

As for other work for the class when at Low Moor, 46117 *Welsh Guardsman* and 46145 *The Duke of Wellington's Regiment* (West Riding) were observed on the Heally Mills–Edge Hill freights, 4.20pm York–Crewe Class Cs, the Congleton–Mirfield parcels and the 11.15pm Huddersfield–Brighouse trips. Low Moor used the locos on any of the shed's diagrams (subject to route restrictions)l, with perhaps the most unusual observation being in April 1962 when 46109 *Royal Engineer* was seen on the heavy Bradford Bridge Street–Huddersfield Hillhouse fitted freight (loaded to 600tons (609,600kg)), which had the assistance of WD 90397 (tender first) from Bradford to Low Moor.

As became common in the mid-1960s, the Low Moor rebuilt Scots appeared on all sorts of trains, having been borrowed by other depots for use, such as:

- 46130 *The West Yorkshire Regiment* on 16 September 1961 on the Liverpool Exchange–Leeds Central express

- 46145 *The Duke of Wellington's Regiment* (West Riding) on 30 December 1961 on a Derby–Bristol parcels
- 46109 *Royal Engineer* on 24 January 1962 on a Glasgow–Manchester express
- 46109 *Royal Engineer* on Whit Monday 1962 on a Halifax–Southport excursion.

Also, the close proximity of the shed too Leeds Holbeck meant that Holbeck could borrow locos at times of shortages, so, for example, on 12 May 1962 46130 *The West Yorkshire Regiment* (Low Moor) worked the 8.50 Leeds–St Pancras express, deputizing for a diesel, and on 10 May 1962 46113 *Cameronian* was used by Leeds Holbeck on the eight-coach express from Leeds to Morecambe.

The stay at Low Moor was relatively short, with two of the locos returned to Leeds Holbeck in the middle of 1962 to cover for diesel failures and the additional summer extras, with the other three engines moving to Mirfield shed in January 1962.

Mirfield

Mirfield shed was formerly 25D, before transfer to the North East Region in September 1956 as shed code 56D. It was a former L&Y/LMS depot between Dewsbury and Huddersfield and had around thirty engines on its books, but was mainly a freight depot with a large allocation of WD 2-8-0s and a smattering of B1 4-6-0s. The only passenger workings were local trains to Huddersfield, Penistone, Bradford and Holmfirth.

The relocation to Mirfield was another step down for the Class7s, as the shed was essentially a freight depot without any of the passenger workings at Low Moor. However, the stay at Mirfield was mercifully short, as the locos were transferred back to Leeds Holbeck in June 1962. At least the proximity of Mirfield to Leeds enabled Holbeck to borrow back some of its old locos for passenger workings. 46145 *The Duke of Wellington's Regiment* (West Riding) was unusual in that it received an overhaul at Crewe Works in April 1962, whilst working from Mirfield.

Some of the workings whilst at Mirfield include 46113 *Cameronian* on 21 March 1962 and on 22 March the engine was also observed on the York–Manchester

46113 Cameronian *(rebuilt December 1950) is seen at Mirfield shed, where it had been allocated in January 1962, but had previously been a long-term resident of Leeds Holbeck, before a move to Low Moor in November 1961 and then to Mirfield. The duties at Mirfield were somewhat humble, but the engine was regularly borrowed by Holbeck and was seen on expresses from Leeds. The engine returned to Holbeck in June 1962, before withdrawal in December 1962 and scrapping at Crewe Works in June 1963. The engine still has the older style of emblem on the tender, which it retained until withdrawal, and has been fitted with AWS (date unknown) and a speedometer drive (February 1961).*

Allocations of Rebuilt Class 7s to Mirfield

Loco		From	Date	To	Date
46113	*Cameronian*	Low Moor	January 1962	Holbeck	June 1962
46117	*Welsh Guardsman*	Low Moor	January 1962	Holbeck	June 1962
46145	*The Duke of Wellington's*	Low Moor	January 1962	Holbeck	June 1962
	Regiment (West Riding)				Received an overhaul at Crewe Works, April 1962.

Red Bank ECS. 46117 *Welsh Guardsman* was seen on 3 June 1962 on a Pontefract–Scarborough excursion.

Saltley

Saltley, coded 21A, was suddenly the beneficiary of thirteen rebuilt Class 7s (and a small number of Jubilees), when most of the rebuilt Class 7s allocated to the Midland Main Line (depots at Nottingham, Kentish Town and Trafford Park) arrived at the depot. The mass transfer coincided with the introduction of the Sulzer Class 4 (Class 45), which was meant to replace steam on the Midland Main Line expresses. George Harrison, a driver at Saltley and later a loco

inspector, said that the rebuilt Class 7s were 'dumped' at Saltley as 'surplus to requirements' and that the depot acted as a 'pool' from which other depots could borrow the engines.

Saltley was a very large depot, with 191 locomotives allocated there in 1954. It had a large allocation of Crab 2-6-0s, Black Five 4-6-0s and 8F 2-8-0s, as well as a stud of 9F 2-10-0s that the Saltley crews absolutely adored.

In May of 1962, two rebuilt Scots, 46106 *Gordon Highlander* and 46118 *Royal Welch Fusilier* from Saltley were reallocated to Leicester for working the heavy Newcastle–Bournemouth train between Leicester and Banbury. However, the use of the rebuilt Class 7s on this train only lasted a few weeks, as it was decided

46141 The North Staffordshire Regiment *(rebuilt October 1950) is seen after its transfer to 21A Saltley in June 1961 on an unfitted freight of 16ton coal wagons. Saltley was a freight depot, so the thirteen rebuilt Class 7s that arrived in the summer of 1961 were put on whatever duties the depot had. One engine, 46162, was observed working the colliery shunts in the Burton area in early 1962. The engines seem to have been borrowed by other depots on numerous occasions before they were all transferred to Carlisle sheds in the summer of 1962.*

46157 The Royal Artilleryman *(rebuilt January 1946) is seen on a fitted freight at Banbury (with the loco depot in the right background) in 1962 when the engine was allocated to 21A Saltley, where they would be seen more on freight trains than on passenger workings.*

that the English Electric Type 3, which brought the train into Leicester from Sheffield, would continue to Banbury and return to Sheffield on the balancing working. The Scots were then transferred back to Saltley, although they did stay long enough to have 15C shed plates placed on the smokebox.

Saltley was primarily a freight depot with workings all over the country, with the most well known being the express freights from Washford Heath or Water Orton to Carlisle, which were the preserve of the three stoker-fitted 9Fs allocated to the depot. The railway press at the time commented that the rebuilt Class 7s would work summer extra trains, thereby reducing the need to use 9F 2-10-0s on these. Certainly, the author saw 46118 *Royal Welch Fusilier* working an express passenger train in the early summer of 1962 at Birmingham New Street and the same engine was observed at Crewe in April 1962 on an express

(reporting number IZ 63). 46137 *The Prince of Wales's Volunteers* (South Lancashire) was seen heading north through King's Norton with an express and 46157 *The Royal Artilleryman* was observed at Camden being serviced, having worked on a passenger train to Euston. 46103 *Royal Scots Fusilier* was seen at York on an express a long way from home, but this illustrates the diverse nature of the trains using Saltley's express power.

However, the rebuilt Class 7s also seem to have been borrowed by lots of depots for various workings, with 46157 *The Royal Artilleryman* observed at Worcester shed. Bristol Barrow Road shed was well known to the Saltley crews for borrowing its Class 7 rebuilds. Saltley put the rebuilt Class 7s into its stud and used them on whatever services the depot had to run, so they were seen on fitted freights to Sheffield, Bradford, Leeds, Bristol and Banbury, as well as being used on

Allocations of Rebuilt Class 7s to Saltley

Loco	From	Date	To	Date
45532 Illustrious	Nottingham	June 1961	Upperby	June 1962
45540 Sir Robert Turnbull	Trafford Park	June 1961	Upperby	June 1962
46103 Royal Scots Fusilier	Kentish Town	June 1961	Upperby	June 1962
46106 Gordon Highlander	Trafford Park	September 1961	Leicester GC	May 1962
46106 Gordon Highlander	Leicester GC	June 1962	Upperby	June 1962
46118 Royal Welch Fusilier	Derby	August 1961	Leicester GC	May 1962
46118 Royal Welch Fusilier	Leicester GC	June 1962	Upperby	July 1962
46122 Royal Ulster Rifleman	Trafford Park	June 1961	Upperby	June 1962
46123 Royal Irish Fusilier	Kentish Town	June 1961	Upperby	June 1962
46132 The King's Regiment Liverpool	Kentish Town	October 1961	Upperby	June 1962
46137 The Prince of Wales's Volunteers (South Lancashire)	Trafford Park	August 1961	Upperby	June 1962
46141 The North Staffordshire Regiment	Trafford Park	June 1961	Upperby	July 1962
46157 The Royal Artilleryman	Nottingham	June 1961	Upperby	July 1962
46160 Queen Victoria's Rifleman	Kentish Town	June 1961	Upperby	July 1962
46162 Queen's Westminster Rifleman	Kentish Town	June 1961	Upperby	July 1962

unfitted loose coupled freights, including coal trains. Perhaps the lowest point was when 46162 *Queen's Westminster Rifleman* was borrowed by 17B Burton shed for a few days and was used on local coal trains and shunted the sidings of Cadley Hill and Donisthorpe collieries, the only time a rebuilt Scot was observed on the Swadlincote Loop. A photograph was taken at Donisthorpe sidings, but the quality is not good enough for publication.

Certainly, the reliability of the Sulzer 4s was not good when introduced in 1961 and failures were common, necessitating the borrowing of some Class 7 power from the nearest depot that had any. Inspector Harrison recollects that a number of the Saltley rebuilt Class 7s were borrowed by Llandudno Junction for summer extras for a few weeks.

Bizarrely, two of Saltley's rebuilt Class 7s received an overhaul at Crewe Works, 46157 *The Royal Artilleryman* in May 1961 and 46132 *The King's Regiment* in April 1962. The rebuilds at Saltley suffered a second mass transfer when they were all moved away in June and July 1962 to Carlisle Upperby, where a number went straight into store, some never to work again.

Annesley

This was an ex-Great Central shed north of Nottingham and its principal workings were the 'windcutter' fast coal trains to London, with 9Fs as well as coal traffic from the many local collieries. The shed had some passenger diagrams, but these were covered by its Black Fives.

However, with the work on the West Coast Main Line to Euston to electrify the route, a number of trains were diverted, one of which was a Euston–Manchester sleeper service. This was now routed from Euston to Bletchley and then cut across to the Great Central Main Line at Calvert. The train then worked north via Leicester and Nottingham and on to Manchester. From Manchester to Nottingham, the train was worked by a Newton Heath EE Type 4 (Class 40), but from Nottingham

46158 The Loyal Regiment (rebuilt September 1952) is seen at Nottingham Victoria after being allocated to 16D Annesley shed in October 1962, where it remained until withdrawal in November 1963 and scrapping in the same month at Crewe Works. The engine had previously been allocated to Trafford Park in December 1960, when it was one of a batch of rebuilt Class 7s allocated to the Midland Main Line, where it shared duties with engines allocated to Nottingham and Kentish Town. The engine is fitted with AWS (January 1959) and a speedometer drive (July 1960) and has acquired the overhead warning flashes applied around 1960–1. The engine also has a welded tender. COLOUR-RAIL

46163 Civil Service Rifleman (rebuilt October 1953) is seen on 13 January 1963 waiting to go off Annesley shed to work the 5.15pm train to Marylebone from Nottingham Victoria. The engine had arrived from Willesden in January 1963 and stayed at Annesley until withdrawal in September 1964. It remained stored at the depot until it began the long journey in January 1965 to Birds Scrapyard at Risca (South Wales), arriving in February and being scrapped the same month. The engine in its prime was a regular on the North Wales coast expresses, being allocated to Holyhead (three times) and Llandudno Junction between April 1960 and its arrival at Willesden in June 1962. The engine is fitted with both AWS (October 1959) and a speedometer drive (February 1962). MALCOLM CASTELDINE

46111 Royal Fusilier (rebuilt October 1947) is seen in April 1963 with a very full coal load in the tender and passing New Basford carriage sidings, where it will pick up its stock and take it down to Nottingham Victoria to work a train to Marylebone. The WD coming in the other direction is 90438 with an empty coal train. The loco had arrived at Annesley in January 1963 from 1A Willesden, but suffered a burst main steam pipe in September 1963, causing a blowback into the cab and injuring the crew. The locomotive was immediately withdrawn and sent to Crewe Works, where it was scrapped in November 1963. MALCOLM CASTELDINE

46112 Sherwood Forester (rebuilt September 1943) sits at the side of Annesley depot in September 1964 and shows off the accident damage caused when it ran into another light engine (a Black Five), which led to its withdrawal in May 1964. After a period stored at the depot, the engine was towed to Cashmores Yard at Tipton shortly after the photograph was taken in September 1964.

Allocations of Rebuilt Class 7s to Annesley

Loco	From	Date	To	Date	Comments
45529 *Stephenson*	Willesden	6 October 1963	Withdrawn	22 February 1964	Stored Willesden, December 1963 until February 1964.
45735 *Comet*	Willesden	6 October 1963	Withdrawn	March/Oct 1964	Stored at Annesley, September 1964 to December 1964 and scrapped at Cashmores, Tipton, January 1965.
46101 *Royal Scots Grey*	Willesden	5 January 1963	Withdrawn	September 1963	Withdrawn following damage to one of the outside cylinders due to water carry over. Stored Willesden, August 1863 to March 1964 and seen at Barrow Hill, 10 April 1964 and scrapped at Slag Reduction, Rotherham.
46111 *Royal Fusilier*	Willesden	12 January 1963	Withdrawn	28 September 1963	Main steam pipe fractured in service in September 1963, resulting in withdrawal.
46112 *Sherwood Forester*	Trafford Park	29 February 1962	Withdrawn	9 May 1964	Accident damage to the front end caused its withdrawal. Stored May 1964 to August 1964 and scrapped Cashmores, Tipton, September 1964.
46114 *Coldstream Guardsman*	Willesden	28 September 1963	Withdrawn	September 1963	Loco arrived at Annesley, but was withdrawn immediately due to poor condition.
46122 *The Royal Ulster Rifleman*	Willesden	29 December 1962	Carlisle Upperby	October 1964	
46125 *3rd Carabinier*	Willesden	21 September 1963	Withdrawn	October 1964	Stored October 1964 to December 1964, before being scrapped at Cashmores, Tipton, January 1965.
46126 *Royal Army Service Corps*	Willesden	1 December 1962	Withdrawn	October 1963	
46143 *The South Staffordshire Regiment*	Trafford Park	29 September 1962	Withdrawn	December 1963	Dropped its inside motion on 7 November 1963.
46153 *The Royal Dragoon*	Trafford Park	29 September 1962	Withdrawn	December 1962	
46156 *The South Wales Borderer*	Willesden	6 October 1963	Withdrawn	10 October 1964	Stored at Annesley for three months, before going to Draper's in Hull for scrapping in January 1965.
46158 *The Loyal Regiment*	Trafford Park	29 September 1962	Withdrawn	19 October 1963	
46163 *Civil Service Rifleman*	Willesden	12 January 1963	Withdrawn	28 August 1964	Big end went in service.
46165 *The Ranger (12th London Regiment)*	Crewe North	29 February 1964	Withdrawn	21 November 1964	The last rebuilt Class 7 at Annesley. Had received an overhaul at Crewe Works at the end of 1962. Was a paper transfer back to Crewe, but was withdrawn from Annesley. Stored at Annesley, November 1964 to February 1965 and scrapped at Wards, Beighton, in March 1965.
46167 *The Hertfordshire Regiment*	Willesden	21 September 1963	Withdrawn	April 1962	

46143 The South Staffordshire Regiment (rebuilt June 1949) is seen shortly after arriving at 16B Annesley shed in October 1962 from 9E Trafford Park and still carries its nameplates. Later in its stay at Annesley it acquired a reputation as being the worst rebuilt Scot at the depot. As a result, it was banned from passenger trains and was only to be used if there was nothing else to use. It suffered damage to its inside motion in November 1963 and was withdrawn in December 1963. Standing behind is a green-liveried Standard Class 5, which was a Shrewsbury resident for many years, but later moved to Patricroft. MALCOLM CASTELDINE

The photograph shows two unknown rebuilt Class 7s meeting face to face. One engine has a lamp on the smokebox front, therefore is not in storage or withdrawn. The author bought the slide, but no details were written on it. Looking at the backing plate, the author would hazard a guess that the loco in the foreground is 46165 The Ranger (12th London Regiment) and therefore the scene would be at Annesley. The condition of the engines was typical of the last few years of those in service across the country.

south it was steam. The Annesley crew would work the train to Leicester and then be relieved by Leicester men for the onwards journey. As this train was heavy (fourteen coaches), a number of Britannia Pacifics were initially allocated, but just as the Annesley fitters had got on top of the Britannias, they were transferred away and replaced by rebuilt Class 7s.

As well as the sleeper service, the rebuilt 7s were also regularly used on the 5.15 Nottingham Victoria–Marylebone and the 11.15 Nottingham–Marylebone parcels, then the return parcels to Nottingham, although a Black Five was also capable of working these trains. During the summer timetable, the rebuilds were also seen on holiday extras, so for example in August 1964 45735 *Comet* was seen at Nottingham with a Mablethorpe–Leicester Central express, the engine coming on at Nottingham for the final leg to Leicester.

The extreme winter of 1962–3 led to a number of Annesley's rebuilt Class 7s being borrowed by other depots, so in April 1963 46101 *Royal Scots Grey*, 46111 *Royal Fusilier* and 46158 *The Loyal Regiment* were observed at Nottingham Midland station and in February 1963 Annesley's 46111 *Royal Fusilier* was seen on the 3.20 Nottingham–St Pancras express.

The sleeper diversion was ended on 5 April 1964, when the train reverted to its original route and the

rebuilt Class 7s worked out their days on whatever traffic the depot could find for them. At the end of the summer service in 1964, the remaining rebuilds at the depot were withdrawn .

The locos allocated to Annesley were not always in good condition (many of the ex-Willesden engines had been in store at Devons Road Bow) and the fitting staff's attention was on maintaining the 9Fs for the 'windcutter' trains, followed by the Black Fives, which could run the services that the Scots were running. On many occasions, the crews would book on expecting a rebuilt Scot, but find themselves with a Black Five instead, as the rebuild was not fit to work. 46126 *The Royal Army Service Corps* and 46143 *The South Staffordshire Regiment* were very run down and 46143 had a reputation for using water, but 46156 *The South Wales Borderer*, 46158 *The Loyal Regiment*, 46163 *Civil Service Rifleman*, 46165 *The Ranger* (12th London Regiment) and 46167 *The Hertfordshire Regiment* were in better condition on arrival. 46165 *The Ranger* (12th London Regiment) had been through Crewe Works in October 1962 and probably had not accumulated a high mileage, so was considered the best of the bunch and was the last operational rebuild Class 7 at the depot.

46112 *Sherwood Forester* was seen on the GCR routes and on special trains, so on 15 September 1962 it was

46112 Sherwood Forester (rebuilt September 1943) is seen on 29 September 1962 at New Basford Carriage sidings just north of Nottingham Victoria station, where it would pick up or drop off carriages from workings at Nottingham Victoria. The engine has lost its nameplates and also the backing plate, which happened when the Class 45 diesel received the same name. It has a stencilled '6112' number on the valance near the front of the engine, which would only be applied at Crewe Works for repair purposes. However, the last recorded visit to Crewe is listed as June 1960, so it is highly likely that the engine had been to Crewe sometime early in 1962. MALCOLM CASTELDINE

46111 Royal Fusilier (rebuilt October 1947) is seen bearing its 16B Annesley shed code as it sits at Cricklewood depot. With the running-down of steam services and the closure of Neasden, locomotives working into Marylebone would have to be serviced at Cricklewood, or sometimes Willesden. 46111 had arrived at Annesley in February 1963 and was withdrawn in October 1963, before being scrapped at Crewe Works in November 1963. MIDLAND RAILWAY TRUST

used on a Beeston–Blackpool special and a day later was seen on a Nottingham–Scarborough excursion.

46112 *Sherwood Forester* was withdrawn in May 1964 following an accident that damaged the front of the engine. During the night of 10 March 1964, with Driver Arthur Wilcox and fireman Alan Newbury on the footplate, the engine was working light to Staveley and stopped at Woodhouse for water. Driver Wilcox shouted to fireman Newbury that they had been given the signal, so the water bag was dropped and 46112 moved forwards at low speed, then bang – they had collided with the tender of another light engine, a Black Five that was not displaying a lamp on the back of the tender. Fireman Alan Newbury was knocked out, slept for twenty-four hours, but then went back to work! As can be seen from photographs, the damage to 46112 would have required a visit to the Works and by this date that would not happen, so the engine was withdrawn and after a period in storage was towed to Tipton for disposal at Cashmores Yard.

The majority of the rebuilt Class 7s when they arrived at the depot already had their nameplates removed and, if not, they were taken off soon after. The best Scot, according to the Annesley staff, was 46165 *The Ranger (12th London Regiment)*, although as with all the Scots at Annesley its riding was lively, to say the least. It had

received an overhaul at Crewe Works in November 1962, so its condition would be better than the usual worn-out units offered up by Willesden. 46143 *South Staffordshire Regiment* was considered the worst Scot at Annesley and was banned from passenger trains and was only turned out from the shed if nothing else was available. It was observed on a coal train near Sheffield just before withdrawal. The sleeper train was actually crewed by Leicester Central crews, who were of the opinion that the rebuilds would have been in better condition if maintained at Leicester shed, as, when the engines first arrived, the sleeper train ran to time or early.

The biggest complaint that the ex-LNER men had with the rebuilt 7s (and to a lesser degree the Stanier Black Fives) was the continual problems with the injectors. As an example, 46163 *Civil Service Rifleman* failed on the sleeper train one evening, with both injectors not working but in possession of a full head of steam. Fireman Newbury said that you were never short of steam on a rebuilt Class 7, but that on one occasion when working an engine hard, for every six shovels he put on the fire four went up the chimney. They were not popular for disposal at the end of the day, with the very long firebox (a 10ft [3m]-long grate) being a difficult job. Generally speaking, the Annesley crews would borrow a V2 2-6-2 wherever possible over a rebuilt Scot.

CHAPTER EIGHT

MAINTENANCE OF THE LOCOMOTIVES

Repair Locations

The LMS created a mileage-based maintenance regime that was followed by BR. Certain types of maintenance were carried out at set mileage or set intervals, the objective being to ensure that locomotives did not fail between examinations. A system of 'garage' or concentration depots (usually the A sheds in each district, so, for example, 9A Longsight for the Manchester districts) was set up to carry out the more complex repairs, but which did not require Works attention. These concentration depots would have a wheel lathe, axle-box lathe and drilling machines, along with facilities for a coppersmith and a tinsmith for white-metal bearings. These 'garage' depots were also generally fitted with wheel drops, so that wheel sets could be dropped out and repairs to axle boxes and journals could be carried out. In the case of a breakdown, components could be sent to Works for repair, returned to the depot and placed back on the engine.

Depots in the district would carry out the necessary mileage examinations and some depots would carry out boiler repairs, including retubing and replacing boiler stays. If more complex work was required, the engine could be repaired at the concentration depot, or sent to Works. Examinations would take place regularly to a defined checklist and also the 'X' day examinations. The 'X' would be the day that a locomotive would be stopped for routine examination, which for express passenger engines would usually be weekly, and on the 'X' day other examinations due, such as a V&P or a boiler washout, would also be carried out. Repairs such as Valve and Piston (V&P) would be car-

ried out at the depots; here, the valves would be removed and the piston rings taken off and any build-up of carbon removed and new piston and valve rings fitted. The blast pipe would also be removed and any carbon build-up cleaned off. In addition, the motion would be taken down and examined and, if necessary, bushes remetalled.

46115 Scots Guardsman *during preservation spent sometime at Tysley Works in the 1990s and this view shows the locomotive with the boiler removed, revealing the saddle for the smokebox and the massive boiler support bracket, which also acted as a frame stay. The locomotive had been initially restored at Dinting, but a major overhaul was required before steaming again. Following its arrival at Tysley, little work was carried out and it moved to Crewe and then Carnforth, where its overhaul was completed.*

The V&P was usually done at 30,000-mile (48,270km) intervals, which, given the intensive working of London Midland Region Class 7 and 8 engines, would occur every twelve to fourteen weeks. The blast pipe, blower ring and exhaust pipe from the ejector were examined every three to five weeks, which would correspond with every third boiler washout. Locomotives would also be given major inspections at the A shed every 40,000–48,000 miles (64,360–77,230km).

Engines placed into store had the motion greased and bags placed over the chimney and safety valves. The engine would also have to be moved a complete wheel revolution from time to time to prevent it becoming stiff. Once a loco was stored serviceable, it could not be withdrawn for six months.

Works Maintenance Regimes

Works visits were governed again by mileage and were categorized as follows below.

Heavy General (HG)

A Heavy General would involve the removal of the boiler, which would then go to the boiler shop. The

46151 The Royal Horse Guardsman (rebuilt April 1953) is seen undergoing a 'P&V' examination at its home depot of 41C Millhouses (Sheffield). This was a regular Piston and Valve examination and would mean the extraction of the three piston valves and pistons, checking for carbon build-up in the valve rings and measuring bore wear. If necessary, the piston and valve rings would be replaced and possibly a rebore of the piston and valve chests would take place. These P&V examinations were done on a mileage basis and would coincide with other maintenance such as boiler washouts.

46151 The Royal Horse Guardsman (rebuilt April 1953). One of the last Royal Scots to be rebuilt, 46151 is seen at Crewe depot with the leading driving axle removed and showing the revised 'J'-type spring hangers fitted when the engines were rebuilt. The 'concentration' depots such as Crewe North had wheel-drop facilities to enable any of the locomotive or tender wheels to be dropped out, axle boxes and journals examined and repaired as required.

All of the rebuilt Class 7s were maintained at Crewe Works and this 1953 photograph shows on the left that the second engine is 46152 The King's Dragoon Guardsman *with its smokebox door open. In the right foreground, 46146* The Rifle Brigade *heads a line of engines including 46527* City of Salford *and further back can be seen an unidentified rebuilt Class 7 (the nameplate has a crest above the name). With the maintenance cycles, both 46146 and 46152 were seen at Crewe Works at the same time for a number of years after this photo was taken.* LNWR SOCIETY

boiler refitted to the engine would not necessarily be the one previously used, as Crewe Works had nine 'spare' 2A boilers, as repairing a boiler would take much longer than the engine overhaul. So, as an example, 46124 *London Scottish* was fitted with six different boilers between rebuilding in December 1943 and withdrawal in December 1962. Most engines had at least three boiler changes, with some having four or five. The most was the seven used on 46112 *Sherwood Forester*. The last locomotive rebuilt, 46156 *The South Wales Borderer*, was not rebuilt until June 1954, but still managed three boiler changes before withdrawal in October 1964.

The tender would be sent to the tender shop for any necessary attention. Again, the tender that came back to the locomotive would not necessarily be the one that was fitted when the locomotive entered Works.

Heavy Intermediate (HI)

The boiler would remain in situ on the loco, but the wheels and motion would be removed and refurbished and the repair would concentrate on the locomotive. Boiler repairs that did not require the boiler to be removed could be carried out, such as retubing. The tender would again go to the tender shop. After an intermediate repair, the engine was fit to run for approximately the average mileage for its class until the next scheduled general repair.

Light Intermediate (LI)

This could apply to just the engine and mechanical parts being repaired as necessary.

45530 Sir Frank Ree *(rebuilt October 1946) is seen at Crewe Works receiving what looks like a Heavy Intermediate overhaul, which meant that the boiler would not be removed from the frames. The date of the photo took some detective work, as according to the EHC the last Works entry was a Light Casual between June 1960 and August 1960. However,* The Railway Observer *notes its last repair as between March 1962 and April 1962, but the chalk mark on the smokebox top to 'cut off' the top lamp bracket is a modification that was only applied to engines from 1963 onwards.*

The Black Five in the background is 45416, which received a Heavy Intermediate overhaul at Crewe between 26 August 1963 and 24 September 1963, so that gives us a good idea of the date. If so, this must have been one of the last visits to Crewe by a rebuilt Class 7. The railway got value for money from the repair, as the loco lasted until December 1965. This engine and 46115 Scots Guardsman *were the last active members of their respective classes.* RAILONLINE

Light Classified (LC)

This would be the repair or replacement of parts that were defective due to failure or accident and would be carried out in order to make the engine serviceable. A casual repair would not have any impact on the main service cycle for the locomotive.

Non-Classified (NC)

This could be for a repair that a shed could not perform, but did not fit into the normal schedule of LI, HI and HG, for example fixing something after a more major repair. It is common on the Engine History Cards to see NC (Rect) (EO), which would translate to Non-Classified (Rectification) (Engine Only). In the case of 46137 *The Prince of Wales's Volunteers* (South Lancashire), it had a Heavy General

between 17 April 1958 and 20 June 1958, then required a NC (Rect) (EO) between 1 July 1958 and 10 July 1958. So the assumption here is that after running-in for a few days, the engine required further attention and rectification.

Examination of the rebuilt Class 7 Engine History Cards shows HG visits taking place every three years, with HI and LI taking place in-between, resulting in the engines visiting Works almost on a twelve-month basis.

The rebuilt Class 7s were all maintained at Crewe Works, where the rebuilding had taken place. It made sense to have all major repairs at Crewe, as the pool of nine spare boilers could be concentrated at one Works. With larger classes, such as Jubilees, the stock of spare boilers was bigger and could be dispersed to Crewe, Derby and St Rollox.

The author has tried to find examples of rebuilt Class 7s being repaired at other workshops, but with-

46132 The King's Regiment Liverpool (rebuilt October 1943) and 46145 The Duke of Wellington's Regiment (West Riding) (rebuilt January 1944) are seen in the sun at Crewe Works on 8 April 1962, after receiving repairs and repainting. 46132 has a Saltley shed plate and was moved to Upperby shortly after leaving Crewe Works. 46145 was at the time a Mirfield engine, although the shed plate attached is 16A Nottingham, to which it was not in fact allocated. BR did not get value from repairing 46145, as it was withdrawn a few months later in November 1962 from Leeds Holbeck. RAILONLINE

out any success. The only example he has found is the repainting of 45531 *Sir Frederick Harrison* at Derby Works in May 1948, when it received its experimental light green livery. However, it is likely that Non-Classified repairs could have taken place at other Works to reduce mileage for 'one-off' repair work. It is also possible that 46120 *Royal Inniskilling Fusilier*, when it was based at Derby and was being tested regarding rough riding, had some changes to bogie spring rates and driving-wheel springs carried out at Derby and was then retested, although the author has found no proof for this.

Final Visits to Works

The Engine History Cards do not record visits to Crewe Works after the end of 1961, but other evidence suggests that they took place after that date, with 46132 *The King's Regiment Liverpool* and 46145 *The Duke of Wellington's Regiment* (West Riding) being photographed in April 1962 having full repaints, suggesting at least an intermediate overhaul. 46152 *The King's Dragoon Guardsman* was in Crewe Works in March 1962, as it was recorded receiving a modified inside connecting rod with roller bearings (the last so fitted) and the same engine appeared again at

Crewe during 1963, having its top smokebox lamp bracket moved.

Because of the continued unreliability of both Peaks and Class 40s, 45531 *Sir Frederick Harrison* was given an overhaul in April 1963 at Crewe Works, when a decision to withdraw the engine was reversed. The engine at this time was allocated to 8A Edge Hill and the shed was still providing steam power on the route to Leeds to cover diesel failure. At this point, a considerable number of Class 7 rebuilds had been withdrawn, although other rebuilt 7s also appear to have had intermediate overhauls at Crewe in 1963. 45530 *Sir Frank Ree* went into Works for what appeared to be a Heavy Intermediate in August–September, as it was photographed there with its wheels and motion removed, but with the boiler still on the frames. 46148 *The Manchester Regiment* was seen freshly repainted ex-Works at Chester in April 1963 and 46165 *The Ranger* (12th London Regiment) was also seen ex-Works at Chester in November 1962, with the loco number stencilled on all the wheels and tender frame. Due to the initial unreliability of the diesels, Class 7 rebuilds could be seen anywhere doing any sort of work, so Mike Bentley (as a Buxton fireman) found himself working on 46152 *The King's Dragoon Guardsman* (a 6J Holyhead engine) on the Whalley Bridge shunt in the Peak district in 1964.

46164 The Artists Rifleman (rebuilt June 1951) is seen at Crewe works, sometime in 1959 to 1961 as it is fitted with AWS and the loco's last recorded visit was to Crewe in 1961. The engine was withdrawn from Sheffield Darnell in December 1962 and scrapped at Crewe works in March 1963. The chalk line applied by the painter to get the cab side numbers in line can be clearly seen.

46140 The King's Royal Rifle Corps (rebuilt May 1936) is seen in the gloomy interior of 8A Longsight shed on 22 September 1963, where it is receiving attention to the motion, with both connecting rods in front of the engine. This shows the very basic working conditions that fitters and shed staff had to work in when maintaining steam locomotives. The engine had arrived at Longsight from across Manchester (Newton Heath) in June 1963 and remained there until a move to Carlisle Upperby in September 1964. There was little regular work at Longsight by 1963 and the engines could be seen on all sorts of workings. RAILONLINE

46111 *Royal Fusilier* had a potentially serious failure when one of the driving-wheel axles broke during service, but fortunately the crew realized something was amiss so the engine was removed from the train and taken to the shed. At this point, the wheel appeared to be still attached to the axle, but when the locomotive was put on the wheel drop and the wheel set lowered down the driving wheel dropped off, narrowly missing one of the shed fitters. The failed wheel set and axles were sent to Crewe and a replacement set of wheels and axle was sent from Crewe in a five-plank wagon. Getting the wheel set out of the wagon presented the shed staff with a problem, but it was achieved with ropes and block and tackle.

CHAPTER NINE

WITHDRAWAL AND DISPOSAL

Withdrawal

Steam locomotives in the early 1960s were withdrawn for three reasons: as part of the programme to replace them with diesels; if they failed while in service; or if they required a repair that proved to be too expensive.

As part of the withdrawal programme as services were replaced with diesel locomotives, sheds would receive a directive to withdraw a particular locomotive, sometimes without any consideration of its condition. For the LM Region, these instructions would come from the Traction Office at Crewe; other regions would have a similar Traction Office that would send out the instructions to withdraw. The 41C Darnell boiler smiths were somewhat surprised to learn that on the day they finished some major boiler repairs on 45536 *Private W. Wood, V.C.*, the depot was instructed by the NE Region to withdraw the engine. This was prob-

ably the NE region getting rid of its rebuilds without thinking to transfer them back to the LM Region. Similarly, 45522 *Prestatyn*, when allocated to 9A Longsight but working from Buxton, had been retubed at Buxton but was withdrawn almost immediately.

When the rebuilt Class 7s failed in service or needed a repair, they were sent to Works for attention and upon inspection if the repairs were considered too expensive, the locomotive was withdrawn while on Works. The first rebuilt Class 7 to be withdrawn was 45514 *Holyhead*, which was at Crewe Works in May 1961 for an overhaul. After the stripping-down process, the frames were found to be badly cracked and it was decided to withdraw the locomotive.

Rebuilt Class 7s fitting the first category included the mass withdrawal of twenty-three locomotives in December 1962, mainly from the Scottish depots of Polmadie (66A) and Corkerhill (67A). The non-stan-

46111 Royal Fusilier (rebuilt October 1947) is seen stored at Devons Road Bow, along with other rebuilt Class 7s (the one behind being a rebuilt Patriot or Jubilee), all from Willesden. During the 1960s, it was common for Willesden to store a number of its rebuilt Class 7s over the winter before the engines would be restored to traffic for further service. In other cases, locos were stored at Devons Road and then withdrawn. 46111 lived another day by being transferred to 16B Annesley in February 1963 and was withdrawn from there in October 1963.

45532 Illustrious *is seen at Carlisle Upperby shed after being withdrawn in February 1964. The engine had been allocated to Upperby from June 1962 and after withdrawal remained on shed until December 1964, being scrapped at Campbell's Yard in Airdrie in January 1965. The AWS conduit can be seen, but the speedometer drive (fitted January 1961) has been removed, along with coupling rods and motion as well as the nameplates, but it retains the welded tender it acquired upon rebuilding. The driving wheels have holes in the axes. After working on the West Coast Main Line, it was one of the rebuilt Class 7s transferred to the Midland Main Line in 1959, when it moved in November that year to Nottingham, before going to Saltley in June 1961 and then to Carlisle Upperby in June 1962.* G. SHARPE

dard example, 46170 *British Legion*, received an overhaul at Crewe Works in 1961, but it was placed in store at Llandudno Junction shed at the end of the summer services in September 1962 and withdrawn in November 1962. By this time, it had been displaced from Crewe North by diesels.

Examples of locomotives failing in service and then being withdrawn include 46163 *Civil Service Rifleman*, with a big end failing on a Great Central line parcels train near Woodford Halse in August 1964, the loco being retired to Woodford shed and then being withdrawn. 46101 *Royal Scots Grey* was withdrawn following a problem with one of the cylinders that resulted in major damage to one of the outside cylinders. 46111 *Royal Fusilier* (Annesley 16B) was withdrawn immediately following a failure of one of the main steam pipes in the smokebox, which caused a blowback into the cab and injuring the crew. An urgent check of other locomotives allocated to Annesley for potential problems with the steam pipes led to the immediate withdrawal of another one.

There were at least two examples of locomotives having been officially withdrawn, then put back into service for a few weeks – 46106 *Gordon Highlander* and 46127 *Old Contemptibles* were both withdrawn in December 1962 from Carlisle Upperby (and reported as being stored from October 1962), but both were observed in steam after the withdrawal date. 46106 was seen at Cricklewood on 27 January 1963 and

46127 was seen on a freight on the North Wales coast on 27 February 1963. The severe winter of 1962–3 caused many diesel locomotives to fail or their heating boilers to malfunction, so many steam locomotives had to be pressed into service as substitute power.

It was not uncommon for locomotives to be sent to Works for overhaul, but when inspected at the Works (usually at Crewe), the scale of the required repairs would be considered uneconomic. Major cracking of frames or cylinder replacement would be the principal reasons. In the early 1960s, a number of locomotive classes came into this category, including an A4 (60010 *Dominion of Canada*) at Darlington in 1965 (the boiler was condemned and no spare boilers were available at this time), a 9F 92177 in 1964, a Jubilee in 1961 and rebuilt Class 7 45514 *Holyhead* from Sheffield Millhouses, which was condemned at Crewe Works in 1961 with cracked frames.

The problem was that the system was rather haphazard, as locomotives having received a major overhaul were then withdrawn as part of the withdrawal programme, whilst locomotives in much worse condition continued to operate until they wore out. An example was 46170 *British Legion*, which was one of the last rebuilt Class 7s locomotives to receive a Heavy General overhaul in September 1961, but was withdrawn in December 1962, having been stored since September 1962. Rebuilt Class 7s in good condition were withdrawn from Llandudno Junction, whilst Annesley was struggling with worn-out example of the same class.

There did not appear to a process whereby locomotives displaced by diesels could be evaluated and good examples transferred elsewhere and others withdrawn. This particularly applied where some locomotives were in a different region (for example, North East, which included Leeds Holbeck, where some of the rebuilt Scots displaced to Low Moor had recently received Works attention at Crewe).

Disposal

A combination of the introduction of the English Electric and Sulzer Type 4s (Class 40 and Class 45/6) and the availability of the more modern Britannias led to the replacement and withdrawal from service of the rebuilt Class 7s. The picture for their withdrawal was as shown in the table above right.

Much of the data regarding the disposal of the locomotives has come from the 'What Happened to Steam' series written by P.B. Hands and published by Defiant Publications and Steam for Scrap by Atlantic Transport Publishers, as well as references in The Railway Observer, but it is known that the records relating to disposal were not always accurate. For example, sometimes locomotives were bought by one company, then resold to another, or the purchasing company may have had more than one disposal yard and the specific yard cannot be identified. The accurate description of disposal locations has not been helped by the lack of photograph evidence, as many of the yards could not (or would not) provide access to enthusiasts. The level of detail documenting the disposal by private scrapyards was also very variable and even if kept at the time has not survived. Many of the yards are no longer in business and some of them have been absorbed by other companies, compounding the lack of accurate information. Also, some contractors went out of business whilst locomotives were being cut up and the engines would then have to be sold on and moved elsewhere for the final cutting up.

At the time that the rebuilt Class 7s were being withdrawn, the practice was to dispose of surplus locomotives at British Railway Works for the associated region (in most cases), so Midland Region locomotives would normally be scrapped at Crewe, Horwich

Withdrawal of Rebuilt Class 7s

Class	1961	1962	1963	1964	1965
Rebuilt Patriots	1	1	4	8	3
Rebuilt Jubilees			2		
Rebuilt Scots		30	15	21	5
Total	1	31	19	31	8

and Derby. Crewe Works was the usual destination for the LM Region large passenger locomotives and this was the case with the first withdrawals of rebuilt Class 7s, which were disposed of at Crewe in 1962, 1963 and 1964. However, as the rate of withdrawal and disposal of steam locomotives accelerated, a decision was made to sell withdrawn locomotives to private scrapyards for final disposal. It must be remembered that by the end of 1958, BR had more than 16,000 steam locomotives in stock and was still building more, so over a ten-year period it had to dispose of all of these, a task beyond its resources. Discussions between BR and the scrap-metal industry started in 1958 and initially the number of locomotives sold was quite small, but accelerated in the 1960s. The majority of the scrap-metal companies were located in steel-making areas, although the copper inner firebox was actually the most valuable item on a locomotive and the rising price of copper encouraged many smaller yards to bid for locomotives.

Disposal to private yards was done by a competitive tender process, whereby locomotives were listed and private yards would bid to buy them, the locomotives going to the highest bidder. The bidder would also have to pay the cost of transport to his yard, so generally yards were located reasonably close to the locomotive's final depot, although as the number of scrapyards increased, companies would bid for locomotives from more remote locations. So in 1964 and 1965, a number of rebuilt Class 7s were disposed of at various private contractor sites around the country.

The locomotives were very often organized in small 'convoys' to get them from the storage location to the scrapyards, so as an example on 13 January 1965, special working 8Z98 conveyed 45735 Comet and 46125

46139 The Welch Regiment *(rebuilt November 1946) is seen at Crewe Works in May 1963, having arrived there from its last depot, Newton Heath, where it was withdrawn in October 1962 and stored until April 1963. Behind it is another rebuilt Scot, which is unidentified but could have been one of seven rebuilt Scots scrapped at Crewe in May 1963. The smokebox dart has had one of the locking handles removed, which seems to have been done to many of the rebuilds upon withdrawal, presumably to prevent them from being steamed again. The motion has also been removed, which was usually done to ensure no problems when towing. The nameplate remains, although the badge above is missing.*

46149 The Middlesex Regiment *(rebuilt April 1945) is seen on 22 September 1963, a few weeks after withdrawal at Longsight shed. It appears that the engine was part way through a repair to the outside cylinder, as the front cylinder cover has been removed and the coupling and connecting rods have also been taken off. A cross marked on the outside of the cylinder indicates that the fitting staff had found something terminal. The engine was subsequently towed to Crewe Works and scrapped on November 1963.* RAILONLINE

3rd Carabinier, which were towed from Annesley depot to Cashmores Yard at Great Bridge, Tipton, in the West Midlands. On some of these convoys, the locomotive doing the pulling was also withdrawn, being steamed for the last time to take itself and other locomotives for scrapping. From an operational point of view, towing scrap locomotives over long distances was rather a headache for the Operating Department, as not only would a crew need to be found for the locomotive doing the towing (as well as a locomotive sometimes), but a 'rider' would also be required for the last dead locomotive in the convoy. The rider's job in winter was not to be recommended. Chris Ward, a fireman at Annesley, said that the only time he was a rider he froze half to death and was glad he only did it once. Very often, the route taken would not be the most direct, as 'paths' would have to be given for this slow-moving convoy, so heavily used routes would have to be avoided. The speed would also be limited and it was not unknown for one of the dead locomotives to run a hot box, requiring the loco to be taken to the nearest shed for attention before moving on.

In most cases, the locomotives were scrapped at a yard reasonably close to the depot from which the locomotive had been withdrawn (or was stored at), so for example most of the rebuilt Class 7s withdrawn from both Carlisle sheds (Upperby and Kingmoor) were sent to private yards in central Scotland (close to the steel mills at Motherwell), for example 46132 *The King's Regiment Liverpool* from Kingmoor went to West of Scotland Shipbreaking at Troon (near Kilmarnock) and a number of locomotives from Carlisle went to the yards in central Scotland, such as McWilliams at Shettlestone and Motherwell Machinery at Wishaw. There were a number of scrapyards in the greater Glasgow area, with yards at Airdrie, Wishaw, Motherwell, Shettlestone and Coatbridge all being close to steel works. Similarly, the batch of rebuilt Scots withdrawn en masse from the Glasgow sheds of Polmadie (66A) and Corkerhill (67A) in 1962 were all disposed of at McWilliams Yard in Shettlestone (after a long period of storage).

However, local disposal did not always happen and some locomotives travelled large distances to get to their final destination, so in February 1964 45535 *Sir Herbert Walker, K.C.B.* travelled from Carlisle Kingmoor to Nottingham Annesley shed for storage (for six months), before being scrapped at William Rigley Wagon Works at Bulwell, Nottingham, in September 1964. 46148 *The Manchester Regiment* was scrapped at Birds Scrapyard at Morriston, near Swansea, travelling from storage at Birkenhead (although the loco was allocated to Holyhead 6J at the time of withdrawal) and observed at Pontypool Road shed en route to South Wales. 46163 *Civil Service Rifleman* was scrapped at Birds Scrapyard at Risca, near Newport (South Wales), having travelled from storage at Annesley shed.

Following withdrawal, locomotives could spend sometime stored, either at the depot from which they were withdrawn, or at a depot nearby that had space to accommodate them. For example, a number of the locomotives withdrawn from Leeds Holbeck (55A) were stored at the nearby Leeds Neville Hill Depot (55H), before being towed in convoy to Crewe Works. Some locomotives spent many months stored following withdrawal and as an example 46107 *Argyll and Sutherland Highlander* was withdrawn in December 1962, but stored at Glasgow Polmadie (66A) until May 1964, when it was sent to McWilliams of Shettlestone. Similarly, 45522 *Prestatyn*, following withdrawal on September 1964, spent eight months stored at Buxton shed (9L), until finally being scrapped at Central Wagon Co., Ince, Wigan, in June 1965. For some reason, the double chimney was removed whilst it was stored at Buxton. The last rebuild to be scrapped was 46140 *The King's Royal Rifle Corps*, which having been

the penultimate locomotive to be withdrawn in November 1964, remained in store at Kingmoor shed until February 1966, before being cut up at McWilliams, Shettleston, in March 1966.

Disposal Sites

The disposal sites and the engines scrapped at them are listed below.

Crewe Works

45514 *Holyhead*, 45521 *Rhyl*, 45523 *Bangor*, 46103 *Royal Scots Fusilier*, 46106 *Gordon Highlander*, 46108 *Seaforth Highlander*, 46109 *Royal Engineer*, 46111 *Royal Fusilier*, 46113 *Cameronian*, 46116 *Irish Guardsman*, 46119 *Lancashire Fusilier*, 46120 *Royal Inniskilling Fusilier*, 46123 *Royal Irish Fusilier*, 46124 *London Scottish*, 46126 *Royal Army Service Corps*, 46127 *Old Contemptibles*, 46130 *The West Yorkshire Regiment*, 46131 *The Royal Warwickshire Regiment*, 46133 *The Green Howards*, 46134 *The Cheshire Regiment*, 46135

The author has been unable to find a view of a rebuilt Class 7 being cut up at Crewe Works, but here is a view of the scrapping shed at Crewe in 1962, showing in the foreground the middle cylinder block from a Princess Royal. A chalked number 6211 can be seen, indicating that this is 46211 Queen Maude, which was scrapped in April 1962. In the shadows in the background, another Princess can be seen and this is likely to be 46212 Duchess of Kent, which was scrapped at the same time as 46211. The majority of the rebuilt Class 7s were cut up at Crewe, with a total of forty-six locomotives being disposed of at the scrapping shed between 1962 and 1964. The surviving engines in 1964 and 1965 were cut up by private scrap dealers.

The East Lancashire Regiment, 46136 *The Border Regiment*, 46137 *The Prince of Wales's Volunteers* (South Lancashire), 46138 *The London Irish Rifleman*, 46139 *The Welch Regiment*, 46141 *The North Staffordshire Regiment*, 46142 *The York and Lancaster Regiment*, 46143 *The South Staffordshire Regiment*, 46144 *The Honourable Artillery Company*, 46145 *The Duke of Wellington's Regiment* (West Riding), 46146 *The Rifle Brigade*, 46147 *The Northamptonshire Regiment*, 46149 *The Middlesex Regiment*, 46150 *The Life Guardsman*, 46151 *The Royal Horse Guardsman*, 46153 *The Royal Dragoon*, 46154 *The Hussar*, 46157 *The Royal Artilleryman*, 46158 *The Loyal Regiment*, 46159 *The Royal Air Force*, 46161 *King's Own*, 46164 *The Artists' Rifleman*, 46167 *The Hertfordshire Regiment*, 46168 *The Girl Guide*, 46169 *The Boy Scout* and 46170 *British Legion*.

Birds, Morriston, Swansea

46148 *The Manchester Regiment*.

Birds, Risca, near Newport

46163 *Civil Service Rifleman*.

Central Wagon Co., Ince, Wigan

45522 *Prestatyn* and 46129 *The Scottish Horse*.

Cashmores, Great Bridge, Tipton, West Midlands

45735 *Comet*, 46112 *Sherwood Forester* and 46125 *3rd Carabinier*.

Hughes Bolckow, North Blyth

45736 *Phoenix*.

Motherwell Machinery & Scrap. Inshaw Works, Wishaw

45512 *Bunsen*, 45530 *Sir Frank Ree*, 46128 *The Lovat Scouts*, 46152 *The King's Dragoon Guardsman* and 46160 *Queen Victoria's Rifleman*.

G.H. Campbell Ltd, Airdrie

This company had two yards in the Glasgow area, Atlas Works in Airdrie (Lanarkshire) and Bognor Road, Shieldhall, Renfrew. It has not been possible to identify which of these two yards disposed of which locomotives, as locomotives could be destined for one yard yet switched to the other. The yards disposed of 45531 *Sir Frederick Harrison* and 45532 *Illustrious*.

McWilliams, Shettlestone, Glasgow

46102 *Black Watch*, 46104 *Scottish Borderer*, 46105 *Cameron Highlander*, 46107 *Argyll and Sutherland Highlander*, 46110 *Grenadier Guardsman*, 46121 *Highland Light Infantry, City of Glasgow Regiment* and 46140 *The King's Royal Rifle Corps*.

P. Wood Slag Reduction Co., Rotherham

46101 *Royal Scots Grey* and 46114 *Coldstream Guardsman*.

J.N. Connel Ltd Union Works, Coatbridge

45545 *Planet*, 46118 *Royal Welch Fusilier* and 46162 *Queen's Westminster Rifleman*.

West of Scotland Shipbreaking Co., Troon

45527 *Southport*, 46132 *The King's Regiment Liverpool*, 46155 *The Lancer* and 46166 *London Rifle Brigade*.

Wards, Beighton, Sheffield

46165 *The Ranger* (12th London Regiment).

William Rigley and Sons, Forest Wagon Works, Bulwell, Nottingham

45535 *Sir Herbert Walker, K.C.B.*

46118 Royal Welch Fusilier *(rebuilt December 1946) is seen at Hereford shed after working on a special train (3X08) alongside an ex-GWR 2-8-0. The bogie wheels are of the Stanier type, whilst the driving wheels are of the Fowler type and the locomotive is fitted with AWS and a speedometer drive, as well as the overhead warning flashes. The engine had spent six years at Crewe North shed, before being moved a number of times in the space of two years from 1960, when it moved to 16A Nottingham (in January 1960), 21A Saltley (in August 1961) and then Carlisle Upperby (in June 1962), remaining at Upperby until withdrawal June 1964.*

Draper's, Hull

46122 *Royal Ulster Rifleman*.

One locomotive that the location for its scrapping is uncertain is 45526 *Morecambe and Heysham*, which was scrapped at either Wards of Coatbridge or Mclellons, Langloan.

45534 E. Tootal Broadhurst *(rebuilt 31 December 1948) is seen in an official portrait in the early 1950s, when the tender still had the first version of the BR emblem. The rebuilt Patriots were normally paired with a welded tender, but 45534 carried a riveted tender from 1953 to September 1960. The engine is yet to be fitted with AWS or a speedometer.*

46170 British Legion *(rebuilt October 1935) is seen at Wigan on 4 August 1962 on a fitted freight, a bit of a come-down from an express passenger train. By 1962, the English Electric Type 4s (class 40) were in command of most of the express passenger trains on the West Coast Main Line, leaving many of the rebuilt Class 7s on more mundane duties. The engine is allocated to 5A Crewe North and within a few weeks of the photo was moved to Llandudno Junction in September 1962, where it went into storage until withdrawn in December 1962.* COLTAS TRUST

CHAPTER TEN

SUMMARY

How would we summarize the career of the rebuilt Class 7 locomotives? In many ways, they were an outstanding locomotive, bearing comparison with the best of the British express passenger 4-6-0s. They formed the backbone of the London Midland Regions express passenger services on the West Coast Main Line from Euston to Glasgow and the steeply graded route from Leeds to Carlisle and were capable of absorbing considerable abuse. As a group of locomotives, they did not seem to attract attention in the same way as the ex-GWR Kings and Castles and did not regularly have the opportunity to display high-speed running, but their value to the Operating Department cannot be underestimated.

Unlike the North Eastern region, which had a considerable number of Pacifics available (which at times seem to have been used on quite humble services), the London Midland Region had to rely on its big 4-6-0s to be the mainstay of its services. It is true that high-speed running was not the order of the day, but day in and day out the big 4-6-0s were consistently overloaded and still managed to run to time. For example, fireman Jimmy McLelland made mention of the London–Glasgow sleeper that left Carlisle at 4.40am and was regularly loaded to sixteen coaches (610tons [619,760kg]), usually with a Class 7 rebuild at the front.

A regular rebuilt Class 7 duty was the Irish Mail (Euston–Holyhead) and this was usually loaded up to seventeen vehicles, but the rebuilt Class 7s were up to the job, although such a load would inhibit high-speed running. This was not to say that under the right circumstances the rebuilds could not run with the best. For example, driver Peter Johnson of Crewe North depot recounts the time he had 46139 The Welch Reg-

The driver of 46155 The Lancer (rebuilt in March 1948) chats with station staff on the platform of Lancaster station in 1962, with the safety valves lifting. The engine at this time was allocated to Crewe North and would have been a regular on services up to Carlisle. The bracket for the driver's name badge can be seen on the cab side, although there is no photographic evidence that they were ever used. The valance edge curves down to meet the drag beam; on the rebuilt Patriots the valance carried on in a straight line to meet the drag beam.

46141 The North Staffordshire Regiment *(rebuilt October 1950) runs through Rugby with a down express for Perth on 18 August 1958, with what looks like a train of nineteen coaches and a well-coaled tender. This sort of load should have had a Duchess at the front, but it was not unusual for the Operating Department to put a rebuilt Class 7 on instead. The fireman would certainly be working hard on this trip, although loads for a rebuild out of Euston regularly exceeded twelve coaches. The engine was allocated to Carlisle Kingmoor, so was working its way home from London.* PETER GROOM

iment on a Crewe–Carlisle semi-fast with six coaches and after the last stop before Carlisle (Penrith) he tried to reach the magic 100mph (161km/h), but failed by a mere 3mph (4.8km/h).

The availability of the engines was good and the fitting staff knew how to keep on top of them. When being operated on the West Coast Main Line, they were regularly overloaded, with trains of fourteen coaches not

being uncommon. The reality was that the West Coast Main Line needed more Pacifics, but in the absence of any additional Class 8 Pacifics, the rebuilt Class 7s had to fill the space. The loadings on the Midland Main Line of around ten coaches was better for the rebuilds and it was on the Midland Main Line that the fastest runs were achieved. The workings from Leeds to Glasgow via the Settle and Carlisle line also saw the preferred loading of

45534 E.Tootal Broadhurst (rebuilt December 1948) is seen on the turntable at Shrewsbury depot. Rebuilt Class 7s were regular visitors to Shrewsbury shed, as they would hand over to a Western Region locomotive for trains from the north to the south coast.Very often, a locomotive 'resting' at Shrewsbury would be used on a fill-in turn to Stafford and back, before working back to the north. 45534 shows off its riveted tender, which was unusual for the rebuilt Patriots as they would normally possess the welded variety. COLTAS TRUST

ten coaches, but the locomotives had to work the very steep gradients. The 1948 Locomotive Exchange Trials showed that the design was fundamentally good and the 2A boiler was one of the best boilers designed for a 4-6-0 locomotive.

The engines were not seen regularly on Royal Train workings, although 46126 *Royal Army Service Corps* was seen on the Royal Train in the mid-1950s.

They were not perfect, with rough riding being an issue, particularly on the rebuilt Scots, and the injectors were also considered to be a weakness. However,

the crews liked them and Inspector Peter Crawley, who was a fireman at 14B Kentish Town, viewed the Scots as a top locomotive, noting that 'they seemed to have unlimited power'. He also reported the run by 46116 *Irish Guardsman* when it hit 101mph (163km/h) on the Midland Main Line when timed by an engineer visiting Derby Works.

Charlie Brooks, another hard-running 14B Kentish Town driver, claimed that 'the Scots were the best locos to hit the Midland Main Line'. Granville Dobson, a fireman at Low Moor shed during the time that some rebuilt Scots were allocated there, also considered them to be magnificent machines and was quoted as saying 'these were undoubtedly the finest steam engines that crossed my path and were a tribute to Sir William Stanier'. Dobson had many fond memories of 46117 *Welsh Guardsman* and 46145 *The Duke of Wellington's Regiment* (West Riding).

Mike Bentley, a fireman at Buxton shed, fired many members of the class and has nothing but praise for the rebuilt Class 7s, with particular memories of 46158 *The Loyal Regiment*, which in his words 'would pull anything'. The technique for firing the engine was to fill the firebox before leaving Manchester Central and then fire 'little and often'. If the fire was built up on shed, it was important that it was burning through before departure. It was vital to put coal down the sides of the firebox, which with a 10ft (3m) length took some effort and accuracy was required. It took the firemen at Annesley sometime to get used to the length of the firebox. The best driving technique was to drive on the first valve of the regulator and vary the cut-off. Mike Bentley remembers a Crewe North driver who had a little wooden wedge, which, once the loco was on the first valve of the regulator, he would wedge in place and then drive on the cut-off. In Mike's words, you had to 'warm up' a rebuild before giving it some more steam. Blasting out of the station with lots of smoke and steam was not the way to drive these engines.

Crewe North firemen would also use the technique of completely filling the firebox before leav-

46158 The Loyal Regiment *(rebuilt September 1952). A view from the fireman's side of the footplate at Manchester Exchange in April 1961. Mike Bentley, who fired and drove these engines, always maintained that 46158 was the best of the lot, being an excellent steamer that 'rode like a coach' – high praise indeed. 46158 was allocated to 9E Trafford Park from December 1961, until it moved to 16D Annesley in October 1963 and was withdrawn from there in November 1963, before being scrapped at Crewe Works the same month.* COLTAS TRUST

46155 The Lancer *(rebuilt August 1950) is seen near Lancaster on the West Coast Main Line on 25 April 1964 on a parcels train. It was a Crewe North engine at the time (it had been allocated to Crewe at least seven times in its rebuilt career). The engine's final allocation was to Carlisle Kingmoor in September 1964, until withdrawal in November 1964. It was scrapped at West of Scotland Shipbreaking, Troon, in February 1965. Unusually, the smokebox top lamp bracket has not been relocated, a process that started in early 1963 but seems to have been haphazardly applied to the rebuilt Class 7s.* KEN TYLER

ing the shed and the fire would be burning through nicely by the time the engine had to leave with its train. Even the Annesley drivers and firemen, who were brought up on LNER locomotives, had a grudging respect for the locomotives, even if they detested the injectors.

Bushbury shed crews had a preference for the rebuilt Patriots, as the cabs offered better protection in wintry wet weather. They also they felt that the ride was better than on the rebuilt Scots and that the 1ft (305mm) reduction in the cylinder diameter did not seem to have any detrimental effect on the pulling power.

46155 The Lancer (rebuilt August 1950) sits in the fading light at Nottingham Victoria on 19 September 1964, waiting to depart with a returning LCGB special, 'The Pennine Limited'. The engine has been nicely turned out by its home depot of Crewe North, but it was to move away a few weeks later when it went to Carlisle Kingmoor, from where it was withdrawn in early in December 1964. MALCOLM CASTELDINE

46169 The Boy Scout (rebuilt May 1945) is seen on the turntable at Patricroft shed, Manchester. The engine at the time was allocated to 1A Willesden and had probably worked into Manchester and would then be diagrammed for a North Wales working from Manchester to Holyhead and back. Rebuilt Class 7s were regulars at Patricroft working North Wales services. The engine moved to 16B Annesley in February 1963 and was withdrawn shortly after in July 1963. A fine example of a Jim Carter photograph. *RAILONLINE*

45526 Morecambe and Heysham (rebuilt rebuilt February 1947) is seen at Preston on 4 August 1963 on the Saturdays only 10.15 Manchester–Workington express. At the time of the photograph, the engine was allocated to 12B Carlisle Upperby and would remain in traffic until October 1964. The pipe work associated with the AWS can be seen on the edge of the platform valance and the speedometer can also be seen. *D. COUSINS*

CHAPTER ELEVEN

PRESERVATION

46100 *Royal Scot*

46100 *Royal Scot* was withdrawn in October 1962 and was purchased by Billy Butlin to be placed on display at Butlin's Skegness holiday camp. When the locomotive was delivered to the Skegness holiday camp, it was to the accompaniment of a guard of honour by the band of the Royal Scots Guards. The loco was given a cosmetic restoration at Crewe Works between October 1962 and May 1963 and painted in LMS Maroon with gold lettering, a livery it never carried in its rebuilt form.

In 1970, the locomotive was loaned from Butlin's to the Bressingham Steam Museum in Norfolk and in 1988 the ownership of the locomotive moved to Bressingham. The engine ran on the museum's demonstration line, but was eventually withdrawn as it needed a complete overhaul. In 2000, the estimate for the overhaul was £306,000 and a successful application was made to the Lottery Heritage fund, with £221,000 being offered in 2002. The initial restoration was planned to be at Bressingham, but it soon became obvious that the restoration required resources not available to Bressingham and the engine was moved to Southall for the restoration to start. The estimated costs continued to rise, with revised estimates indicating £600,000. A second application was made to the Lottery Heritage fund, resulting in a further £208,000 being granted in 2006. However, the costs have continued to rise, reaching in excess of £1m. One of the reasons for the increased costs was the replacement of volunteers with paid staff and contractors. The boiler was sent to Chatham and the frames went to Tysley.

However, after the restoration was complete and the engine was sent to the West Somerset Railway for its enthusiast weekend, it became clear that something was seriously wrong with the locomotive. The engine was on a low loader being moved when the trailer caught fire, causing serious damage to the bogie and middle cylinder. A decision was made to strip the locomotive to repair the damage caused by the fire and also to identify the poor performance. The loco has been at Crewe, where the wheels have been removed and a major misalignment of the axle boxes has been found and this is to be rectified. In the spring of 2013, there is no announcement of when the engine will be repaired and put back into traffic.

46115 *Scots Guardsman*

Withdrawn from Carlisle Kingmoor in January 1966, 46115 *Scots Guardsman* was the last rebuilt Class 7 in traffic and was purchased by Mr R. A. (Bob) Bill of Bill Switchgear Ltd. After removal from its last depot, the locomotive was stored on the Keighley and Worth Valley Railway, Howarth, before going to the Dinting Railway Centre, Glossop, in May 1969.

Members of the Bahamas Locomotive Society had promised Mr Bill that they would restore the engine to working order, which was probably why they were able to get him to move it from the Worth Valley. Soon after the engine arrived, it was repainted in the LMS 1946 passenger livery – completed in under six months and all achieved outside in the open air, except for the last coat of paint. The accuracy of the livery, once completed, brought praise from David Jenkinson, the noted authority on LMS liveries.

6115 Scots Guardsman. Following preservation, 6115 Scots Guardsman was restored at Dinting by the Bahamas Locomotive Society into its 1946 LMS livery. It was the only loco to receive this livery and carry the smoke deflectors. The engine ran on the main line twice before changes to boiler requirements led to it being used at the Dinting Railway Centre. The well-known Head of Education of the National Railway Museum, the late David Jenkinson (an expert on LMS liveries), gave his approval to the accuracy of the restoration executed at Dinting.

46115 Scots Guardsman (rebuilt August 1947) after withdrawal in December 1965. It is seen following its purchase by Mr Bill and preservation and move from 12A Carlisle Kingmoor to the Keighley and Worth Valley Railway at Howarth, where it is seen in the condition in which it was withdrawn, complete with yellow warning stripe and 12A shed plate. The engine remained for sometime at Howarth in the yard before moving to Dinting Railway Centre, where it was overhauled and repainted in LMS 1946 livery. As far as is known, the only work carried out at Howarth was to clean the engine and motion. The latest restoration carried in 2008 has returned the engine to its BR green livery, but without the yellow stripe.

6115 Scots Guardsman (rebuilt August 1947) is seen at its renaming ceremony at Dinting (where it had been restored) in October 1969, with the band of the Royal Scots Guards in attendance. The renaming was carried out by Lt J.A. Napier of the Regiment.

6115 Scots Guardsman (rebuilt August 1947). Following the long restoration at Dinting, 6115 only ran on the main line twice in its impressive 1946 LMS livery and the loco is seen here on a York–Guide Bridge train on 14 October 1978. Changes to the boiler requirements for main-line steam by BR meant that the boiler would have to be removed for a full internal examination, something that the Bahamas Locomotive Society, who restored the engine, could not afford to do, so the engine only ran at the Dinting site after its two main-line outings.

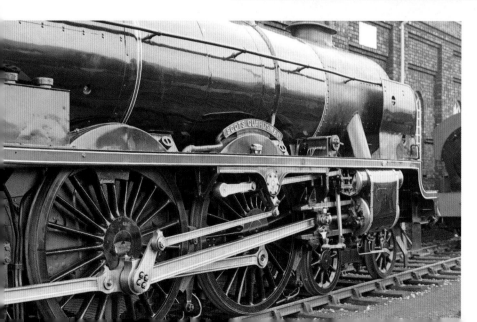

6115 Scots Guardsman at Dinting in the early 1980s showing off its original Fowler driving wheels and the replacement Stanier bogie wheels. The engine is in the LMS 1946 livery and the locomotive has the smoke deflectors and was the only member of the rebuilt Scots to have this combination. The locomotive after withdrawal had been stored at the Keighly and Worth Valley Railway, before moving to Dinting where it was overhauled. It ran only twice on the main line before needing further boiler work.

On 5 October that year, the engine was renamed by Lt J.A. Napier of the Scots Guards and a detachment from the Scots Guards Regiment, who provided the essential pipe band and regalia. There is evidence to state that a quotation was requested from the Hunslet Engine Co. (where the Jubilee Class Bahamas was being overhauled at that time) to have the engine overhauled, and it is assumed this was not pursued because of the cost.

Work on the mechanical overhaul by society members commenced soon after and, following discussion with the local BR inspectors, the boiler was retubed (fire tubes only) and other work done in the firebox to stays and so on. The lubrication system was overhauled, all the motion checked, slide bars reground flat, new tyres fitted to the leading tender wheels and plenty of work was done replating and patching the engine and tender drag boxes. The boiler cladding was extensively repaired and all the brake rigging was overhauled.

Although this might not seem a lot of work by today's standards, it still took eight years to complete – it must be remembered that the Society had no workshop (just the covered accommodation offered by the single-road ex-GCR engine shed and *Scots Guardsman* was a tight fit). The site was still being developed and the Society was running and maintaining all of the other engines and, of course, the restoration was governed by what had to be done to please the BR inspectors – who at that time 'knew a good 'un from a bad 'un!'

The engine had a test run from Guide Bridge to Sheffield prior to its two main-line trips in October and November 1978. The reason for only two trips was the BR ruling, which came into being that year, that any engine to run on the main line was to have a full internal exam of the boiler. Clearly, the Society, only just having retubed and repainted the engine, had no wish to strip it down once more (as well as not having the

6115 Scots Guardsman *(rebuilt August 1947) is seen at Tysley in 1989 and displays the splendid 1946 LMS livery that it carried when rebuilt. It was the first engine to receive the smoke deflectors (fitted in 1947) and was the only engine to carry the 1946 livery and be fitted with deflectors, as the majority were not fitted with deflectors until 1950–2. The engine has had the AWS system removed, as this was out of period for the livery. The engine has subsequently been repainted in its BR passenger green livery since returning to service in 2009.*

46115 Scots Guardsman is seen outside its current home depot of Carnforth in 2008 at an open day at the depot. The engine had been rebuilt at Carnforth and is currently based there for working steam rail tours, usually over the Settle and Carlisle line between Hellifield and Carlisle. ANDREW TOOLEY

46115 Scots Guardsman is seen at Carnforth at an open day in 2008. The speedometer, which was missing when the engine was withdrawn in 1966, has been reinstated and the headboard harks back to the halcyon days of the rebuilt Scots over the Settle and Carlisle line. ANDREW TOOLEY

46115 Scots Guardsman *is seen at Carnforth on 27 August 2008 at an open day. The 'bash plate' to protect the AWS contact shoe at the front of the bogie can be clearly seen.* ANDREW TOOLEY

extra funds). The Boiler Insurance Inspector was quite happy without this procedure, so the Society continued to run the engine at Dinting on brake van rides until the expiry of its boiler certificate, at which point the Society commenced a fund to pay for new tubes.

The engine's only time away from Dinting was the visit to the Liverpool Road Museum in Manchester as part of the Liverpool and Manchester Railway's 150th anniversary festivities in 1980. The engine always went well, although structurally it was somewhat tired. The back end was a bit lively when out on the main line (Pete Skellon prepared a report for the Society's Loco Committee following his footplate trip from Chinley to Sheffield for the November 1978 trip). Pete Skellon has very fond memories of this engine during its time with the Society – and one day hopes to have a trip behind it now that it's back in traffic.

The loco required a complete boiler strip-down and bottom-end overhaul and arrived at Tysley in 1989. A piston and valve examination was carried out and some motion bushes replaced before the locomotive was completely stripped, but then no further work took place. The locomotive then left for Crewe in 1998.

Bob Bill's son Peter sold the engine to the 46115 *Scots Guardsman* Trust, but the loco was then transferred to Waterman Railways, before eventually becoming the property of West Coast Railways, Carnforth, who are the current owners. The Society's Chief Engineer was able to provide the guys at Carnforth with a comprehensive 'state of the engine' report from when the Society had it, which assisted them when it came to planning an overhaul and also provided them with the original smokebox number plate.

The locomotive was restored back into its BR passenger green livery in 2008. On 6 September 2009, the locomotive hauled a special train to commemorate the 100th anniversary of the Girl Guides and was temporarily renumbered and renamed 46168 *The Girl Guide*.

46115 Scots Guardsman (rebuilt August 1947) is seen in 2011 following its restoration and use on the main line. The bevel rim to the bogie wheel indicates that this was a Stanier replacement for the original Fowler wheels. The driving wheels are the original Fowler wheels fitted when the engine was built and retained after rebuilding. *T. COLEMAN*

CLOCKWISE, FROM TOP LEFT:

46115 Scots Guardsman (rebuilt August 1947). Another detail shot of 46115 in 2011 showing the various lubricators on top of the platform and the large supporting bracket under the boiler. T. COLEMAN

46115 Scots Guardsman (rebuilt August 1947). A view of the platform from the driver's side taken in 2011, which not only shows the lubricators on the platform, but also the AWS conduit clipped to the platform edge. T. COLEMAN

46115 Scots Guardsman (rebuilt August 1947). A view inside the cab showing the driver's side. The speedometer can be seen at the bottom left. The box on the upper right is part of the modern warning system that had to be fitted to engines working on the main line. T. COLEMAN

APPENDICES

Appendix 1: Rebuilt Class 7s – Details

Loco Number	Name	Re built	Re numbered	Smoke Deflectors	AWS	Speedometer	Withdrawn	Comments
45512	*Bunson*	26/07/48	31/07/48	14/06/52	27/11/1959	08/02/1961	27/03/1965	
45514	*Holyhead*	26/03/47	22/05/48	30/12/50	07/11/1959	N	27/05/1961	The first rebuilt Class 7 withdrawn. Likely to have received speedometer at works visit in May1961.
45521	*Rhyl*	31/10/46	14/08/48	31/10/49	05/12/1959	31/05/1961	28/09/1963	First Patriot rebuilt.
45522	*Prestatyn*	07/02/49	09/10/48	10/05/52	19/11/1959	11/02/1961	19/09/1964	Last Patriot to be rebuilt.
45523	*Bangor*	08/10/48	09/10/48	25/03/52	13/01/1960	15/06/1961	25/01/1964	
45525	*Colwyn Bay*	20/08/48	21/08/48	05/02/53	09/10/1959	12/11/1960	11/05/1963	
45526	*Morecombe and Heysham*	06/02/47	02/07/49	04/11/50	13/11/1959	17/12/1961	24/10/1964	
45527	*Southport*	13/09/48	18/09/48	23/11/51	07/03/1959	10/09/1960	05/12/1964	
45528	*REME*	23/08/47	12/03/49	23/08/52	29/01/1960	26/08/1961	19/01/1963	Named REME October1959
45529	*Stephenson*	05/07/47	07/08/48	15/09/51	03/10/1959	04/08/1961	22/02/1964	Named *Stephenson* July 1948
45530	*Sir Frank Ree*	19/10/46	10/04/48	11/11/50	24-3-195	19/05/1962	01/01/1966	
45531	*Sir Frederick Harrison*	13/12/47	22/05/48	28/07/51	13/06/1959	03/05/1961	30/10/1965	Painted in experimental light green livery 20 May 1948 and repainted at Derby works.
45532	*Illustrious*	03/07/48	03/07/48	18/08/51	06/11/1959	05/01/1961	01/02/1964	

continued…

Loco Number	Name	Re built	Re numbered	Smoke Deflectors	AWS	Speedometer	Withdrawn	Comments
45534	E Tootal Broadhurst	31/12/48	31/12/48	08/10/53	24/05/1959	17/09/1960	09/05/1964	
45535	Sir Herbert Walker KCB	25/09/48	25/09/48	12/01/52	24/01/1959	17/06/1960	26/10/1963	
45536	Private W C Woods VC	12/11/48	13/11/48	10/12/52	21/02/1959	17/02/1961	29/12/1962	
45540	Sir Robert Turnbull	01/11/47	22/05/48	08/01/52	07/02/1959	05/01/1961	06/04/1963	Painted in experimental light green livery 30 May 1948.
45545	Planet	05/11/48	06/11/48	18/02/52	14/03/1959	14/10/1960	30/05/1964	Named Planet when rebuilt. Fowler driving wheels.
45735	Comet	01/05/42	19/06/48	19/01/52	25/05/1959	19/12/1960	30/10/1964	Retained flat top feed cover until withdrawn.
45736	Phoenix	15/04/42	02/10/48	21/12/50	11/09/1959	26/05/1961	26/09/1964	
46100	Royal Scot	10/06/50	19/06/48	17/06/50	16/05/1959	08/10/1960	13/10/1962	Preserved.
46101	Royal Scots Grey	01/11/45	01/05/48	10-91949	28/02/1959	09/05/1962	31/08/1963	
46102	Black Watch	13/10/49	04/09/49	13/10/49	Y	Y	29/12/1962	
46103	Royal Scots Fusilier	26/06/43	23/10/48	Not on EHC ?	19/06/1962	19/12/1962		Fitted with BTH speedo in 1943.
46104	Scottish Border	01/03/46	07/08/48	11/08/51	Y	Y	29/12/1962	AWS and speedo fitted by 1962
46105	Cameron Highlander	01/03/48	01/05/48	27/12/52	Y	Y	29/12/1962	
46106	Gordon Highlander	01/09/49	12/06/48	08/10/49	31/10/1959?		08/12/1962	Britannia style deflectors fitted July 1954
46107	Argyll and Sutherland Highlander	01/02/50	03/04/48	25/02/50	N	N	29/12/1962	
46108	Seaforth Highlander	21/08/43	08/05/48	25/02/51	28/02/1959	04/11/1961	26/01/1963	
46109	Royal Engineer	21/07/43	01/05/48	Not on EHC	Y	Y	27/12/1962	No date for AWS and SPEEDO.
46110	Grenadier Guardsman	21/01/53	15/05/48	21/02/53	11/07/1959	21/05/1960	22/02/1964	
46111	Royal Fusilier	01/10/47	20/11/48	11/08/51	18/04/1959	18/06/1960	28/09/1963	

continued...

Loco Number	Name	Re built	Re numbered	Smoke Deflectors	AWS	Speedometer	Withdrawn	Comments
46112	Sherwood Forester	14/09/43	11/09/48	08/10/49	N	18/06/1960	09/05/1964	
46113	Cameronian	15/12/50	14/05/49	30/12/50	Y	25/02/1961	17/12/1962	
46114	Coldstream Guardsman	28/06/46	???	09/09/50	16/05/1959	04/11/1961	28/09/1963	
46115	Scots Guardsman	20/08/47	22/01/49	Not on EHC	03/10/1959	05/11/1960	01/01/1966	Preserved and running on the rail network.
46116	Irish Guardsman	10/08/44	25/09/48	04/11/50	11/07/1959	23/04/1960	17/08/1963	
46117	Welsh Guardsman	19/12/43	29/05/48	04/11/50	Y	Y	26/11/1962	No date for AWS and SPEEDO.
46118	Royal Welch Fusilier	17/12/46	12/02/49	22/04/50	20/11/1959	08/10/1960	13/06/1964	
46119	Lancashire Fusilier	02/09/44	31/07/48	10/09/49	11/07/1959	09/09/1961	16/11/1963	
46120	Royal Inniskilling Fusilier	07/11/44	12/06/48	14/07/51	05/09/1959	31/12/1960	06/07/1963	
46121	Highland Light Infantry The City of Glasgow Regiment	13/08/46	09/10/48	20/04/51	N	N	29/12/1962	Never fitted with AWS or Speedometer. Re named from HLI to Highland Light Infantry on January 1949.
46122	Royal Ulster Riflemen	22/09/45	17/04/48	25/03/50	31/10/1959	10/09/1960	17/06/1964	
46123	Royal Irish Fusilier	05/05/49	26/06/48	13/06/53	28/02/1959	18/06/1960	03/11/1962	
46124	London Scottish	31/12/43	10/04/48	31/12/49	16/05/1959	08/06/1960	29/12/1962	
46125	3rd Caribinier	07/08/43	11/09/48	05/11/49	03/10/1959	17/06/1961	03/10/1964	
46126	The Royal Army Service Corp	30/06/45	04/12/48	06/10/51	28/11/1959	31/12/1960	05/10/1963	
46127	The Old Contemptables	02/08/44	01/05/48	11/08/51	02/01/1960	Y	08/12/1962	Speedo seen on photo taken August 1962.
46128	The Lovat Scouts	04/06/46	26/02/49	28/06/51	28/02/1959	21/05/1960	01/05/1965	
46129	The Scottish Horse	31/12/44	26/06/48	17/06/50	08/08/1959	05/11/1960	06/06/1964	
46130	The West Yorkshire Regiment	06/12/49	22/05/48	31/12/49	16/05/1959	05/11/1960	17/12/1962	
46131	The Royal Warwickshire Regiment	14/10/44	07/08/48	15/07/50	N	N	03/11/1962	
46132	The Kings Regiment (Liverpool)	05/11/43	24/04/48	06/10/51	18/04/1959	08/10/1960	01/02/1964	
46133	The Green Howards	01/07/44	29/01/49	15/01/49	N	30/12/1961	23/02/1963	AWS not fitted by 1963.
46134	The Cheshire Regiment	31/12/53	13/11/48	02/01/54	31/10/1959	22/04/1961	01/12/1962	
46135	The East Lancashire Regiment	01/01/47	01/07/48	01/07/48	30/01/1960	09/09/1961	24/12/1962	
46136	The Border Regiment	22/03/50	31/07/48	25/03/50	31/10/1959	28/01/1961	28/03/1964	
46137	The Prince of Wales Volunteers South Lancashire	26/03/55	08/05/48	26/03/55	03/10/1959	28/01/1961	03/11/1962	
46138	The London Irish Riflemen	09/06/44	15/01/49	09/09/50	16/05/1959	31/12/1960	09/02/1963	

continued...

Loco Number	Name	Re built	Re numbered	Smoke Deflectors	AWS	Speedometer	Withdrawn	Comments
46139	The Welch Regiment	12/11/46	22/05/48	29/05/50	18/04/1959	21/05/1960	13/10/1962	Repainted in experimental green livery in 1948.
46140	The Kings Royal Rifle Corp	30/05/52	08/01/49	14/06/52	18/04/1959	18/06/1960	30/10/1965	
46141	The North Staffordshire Regiment	28/10/50	03/07/48	04/11/50	11/07/1959	05/11/1960	18/4/1964	
46142	The York and Lancaster Regiment	14/02/51	03/07/48	24/02/51	05/09/1959	10/09/1960	11/01/1964	
46143	The South Staffordshire Regiment	07/06/49	04/09/48	18/04/53	18/04/1959	21/05/1960	21/12/1963	
46144	Honourable Artillery Company	09/06/45	05/06/48	17/05/52	11/07/1959	13/08/1960	11/01/1964	
46145	The Duke of Wellington's Regiment (West Riding)	29/01/44	02/10/48	27/01/51	Y	Y	03/06/1962	No date for AWS and SPEEDO.
46146	The Rifle Brigade	08/10/43	12/06/48	08/10/49	28/11/1959	05/11/1960	01/12/1962	
46147	The Northampton Regiment	26/09/46	15/01/49	09/02/51	10/09/1960	15/07/1961	01/12/1962	
46148	The Manchester Regiment	09/07/54	26/06/48	17/07/54	16/05/1959	08/10/1960	14/11/1964	
46149	The Middlesex Regiment	17/04/45	24/04/48	03/12/49	18/04/1959	18/06/1960	31/08/1963	
46150	The Life Guardsman	19/12/45	22/01/49	24/02/51	16/05/1959	16/06/1961	23/11/1963	
46151	The Royal Horse Guardsman	16/04/53	23/10/48	18/04/53	13/06/1959	13/08/1960	29/12/1962	
46152	The Kings Dragoon Guardsman	11/08/45	19/06/48	31/12/49	05/09/1959	08/10/1960	17/04/1965	
46153	The Royal Dragoon	19/08/49	20/08/49	10/09/49	21/03/1959	22/04/1961	22/12/1962	
46154	The Hussar	16/03/48	17/04/48	25/10/51	18/04/1959	16/07/1960	01/12/1962	
46155	The Lancer	18/08/50	03/07/48	28/01/50	11/07/1959	15/07/1961	12/12/1964	
46156	The South Wales Borderer	28/05/54	05/12/49	19/06/54	28/11/1959	25/03/1961	10/10/1964	
46157	The Royal Artillerymen	19/01/46	18/12/48	22/04/50	05/09/1959	Y	04/01/1964	Speedo seen on photo taken August 1962
46158	The Loyal Regiment	24/09/52	09/10/48	04/10/52	24/01/1959	16/07/1960	19/10/1963	
46159	The Royal Airforce	13/10/45	18/09/48	15/07/50	21/03/1959	?	01/12/1962	
46160	Queen Victoria's Riflemen	10/02/45	25/09/48	28/01/50	16/05/1959	04/11/1961	01/05/1965	
46161	Kings Own	12/10/46	31/07/48	10/09/49	03/01/1960	19/05/1962	21/11/1962	
46162	Queens Westminster Rifleman	07/01/48	17/04/48	04/11/50	18/04/1959	21/05/1960	30/05/1964	
46163	Civil Service Rifleman	08/10/53	20/11/48	31/10/53	31/10/1959	24/02/1962	29/08/1964	
46164	The Artist Rifleman	23/06/51	17/04/48	28/10/1950	16/05/1959	25/03/1961	29/12/1962	
46165	The Ranger (12th London Regiment)	12/07/52	23/10/48	12/07/52	13/06/1959	13/08/1960	21/11/1964	
46166	The London Rifle Brigade	10/01/45	31/07/48	17/05/52	18/04/1959	27/01/1962	19/09/1964	
46167	The Hertfordshire Regiment	13/12/48	18/12/48	09/08/52	11/07/1959	10/09/1960	11/04/1964	
46168	The Girl Guide	27/04/46	04/09/48	07/04/51	13/06/1959	13/08/1960	02/05/1964	
46169	The Boy Scout	12/05/45	15/05/48	28/01/50	11/07/1959	05/10/1960	25/05/1963	
46170	British Legion	31/10/35	24/04/48	02/12/51	08/04/1959	09/09/1961	01/12/1962	

Regional Allocations

The following eleven tables provide complete listings of all the regional allocations of LMS/BR Class and 4-6-0 Rebuilds, giving full details of the name and number of locomotive, its original provenance, the date on which it first entered service and finally where and when it ended up. There is also a section of comments detailing any known peculiarities in the train's life as well as modifications and other items.

Appendix 2: Leeds Holbeck Allocation

Loco	From	Date	To	Date	Comments
5526 *Morecambe and Heysham*	Edge Hill	June 1947	Edge Hill	July 1947	On loan for one month to Leeds after conversion.
5735 *Comet*	Crewe Works	May 1942	Camden	July 1943	
5736 *Phoenix*	Crewe Works	June 1942	Camden	July 1943	
46103 *Royal Scots Fusilier*	Camden	August 1943	Kentish Town	October 1958	Stored Farnley Junction, January 1963 to August 1963.
46108 *Seaforth Highlander*	Polmadie	September 1943	Longsight	December 1952	
46109 *Royal Engineer*	Crewe Works	August 1943	Low Moor	September 1961	
46109 *Royal Engineer*	Low Moor	June 1962	Withdrawn	December 1962	Stored Neville Hill, January 1963 to December 1963. Scrapped Crewe Works, December 1963.
46112 *Sherwood Forester*	Holyhead	February 1953	Nottingham	November 1959	
46113 *Cameronian*	Crewe North	February 1951	Mirfield	November 1961	
46113 *Cameronian*	Mirfield	June 1962	Withdrawn	December 1962	Stored Neville Hill, January 1963 to May 1963. Scrapped Crewe Works, June 1963.
46117 *Welsh Guardsman*	Camden	March 1944	Low Moor	November 1961	
46117 *Welsh Guardsman*	Mirfield	June 1962	Withdrawn	November 1962	Stored Neville Hill, January 1963 to November 1963. Scrapped Crewe Works, December 1963.
46119 *Lancashire Fusilier*	Crewe North	December 1946	Willesden	January 1947	On load at Leeds.
46120 *Royal Inniskilling Fusilier*	Derby	June 1949	Longsight	January 1950	On load at Leeds.
46125 *3rd Carabinier*	Crewe North	May 1949	Crewe North	August 1949	On loan at Leeds.
46126 *Royal Army* Service Corps	Crewe North	March 1946	Crewe North	May 1946	On loan at Leeds.
46126 *Royal Army* Service Corps	Edge Hill	September 1949	Camden	October 1949	On loan at Leeds.
46126 *Royal Army* Service Corps	Carlisle	March 1955	Carlisle	April 1955	
46128 *The Lovat Scouts*	Camden	August 1946	Crewe	December 1946	
46128 *The Lovat Scouts*	Holyhead	November 1949	Crewe North	January 1950	On loan at Leeds.
46130 *The West Yorkshire Regiment*	Kentish Town	November 1959	Low Moor	September 1961	
46130 *The West Yorkshire*	Low Moor	June 1962	Withdrawn	December	Stored Farnley Junction,

				1962	January 1963 to September 1963.Scrapped Crewe Works, October 1963.
46132 *The King's Regiment Liverpool*	Crewe North	September 1945	Crewe North	October 1945	
46133 *The Green Howards*	Upperby	July 1944	Carlisle	August 1944	On loan at Leeds.
46133 *The Green Howards*		December 1946	Kentish Town	December 1958	
46135 *The East Lancashire Regiment*	Crewe North	August 1962	Withdrawn	December 1962	
46136 *The Border Regiment*	Crewe North	August 1962	Upperby	September 1962	On loan at Leeds.
46141 *The North Staffordshire Regiment*	Carlisle	January 1956	Upperby	January 1956	Only allocated for two weeks.
46145 *The Duke of Wellington's Regiment (West Riding)*	Longsight	January 1953	Low Moor	November 1961	
46145 *The Duke of Wellington's Regiment (West Riding)*	Mirfield	June 1962	Withdrawn	November 1962	Stored Farnley Junction, December 1962 to September 1963. Scrapped Crewe Works, October 1963.
46146 *The Rifle Brigade*	Crewe North	March 1946	Crewe North	April 1946	
46149 *The Middlesex Regiment*	Crewe North	August 1945	Crewe North	September 1945	
46160 *Queen's Victoria's Rifleman*	Crewe North	June 1945	Crewe North	June 1945	On loan for two weeks.
46161 *King's Own*	Crewe North	July 1962	Withdrawn	November 1962	Stored at Neville Hill December 1962 to November 1963. Scrapped Crewe Works, December 1963.
46170 *British Legion*	Crewe North	July 1943	Camden	August 1943	On loan.

Appendix 3: Crewe North Allocation

Loco	From	Date	To	Date	Comments
45514 *Holyhead*	Holyhead	June 1947	Bushbury	July 1947	
45522 *Prestatyn*	Crewe Works	February 1949	Longsight	June 1949	
45523 *Bangor*	Crewe Works	October 1948	Camden	July 1951	
45525 *Colwyn Bay*	Crewe Works	August 1948	Bushbury	October 1949	
45525 *Colwyn Bay*	Upperby	September 1950	Camden	July 1951	
45525 *Colwyn Bay*	Camden	September 1951	Camden	July 1952	
45525 *Colwyn Bay*	Camden	September 1952	Edge Hill	October 1952	On loan to Edge Hill.
45528 *REME*	Bushbury	June 1948	Longsight	October 1948	
45528 *REME*	Longsight	March 1950	Camden	July 1951	
45528 *REME*	Camden	September 1951	Camden	June 1953	
45528 *REME*	Camden	September 1953	Camden	June 1954	

45528 *REME*	Camden	September 1954	Longsight	February 1955	
45528 *REME*	Holyhead	May 1955	Longsight	January 1956	
45528 *REME*	Longsight	April 1956	Holyhead	April 1957	On loan to Holyhead.
45528 *REME*	Holyhead	April 1957	Willesden	January 1962	
45529 *Stevenson*	Bushbury	April 1949	Camdcn	July 1952	
45529 *Stevenson*	Camden	September 1952	Camden	June 1953	
45529 *Stevenson*	Camden	September 1953	Camden	November 1953	
45529 *Stevenson*	Camden	January 1954	Camden	March 1958	
45529 *Stevenson*	Camden	April 1958	Willesden	January 1961	
45534 *E. Tootal Broadhurst*	Crewe Works	December 1948	Bushbury	October 1949	
45534 *E. Tootal Broadhurst*	Longsight	September 1960	Willesden	June 1961	
45534 *E. Tootal Broadhurst*	Llandudno Junction	June 1963	Withdrawn	May 1964	Stored Crewe North, May 1964 and scrapped Crewe Works, June 1964.
45535 *Sir Herbert Walker, K.C.B*	Crewe Works	October 1948	Camden	September 1950	
45535 *Sir Herbert Walker, K.C.B*	Edge Hill	December 1950	Camden	July 1951	
45535 *Sir Herbert Walker, K.C.B*	Camden	September 1951	Edge Hill	October 1952	
45535 *Sir Herbert Walker, K.C.B*	Edge Hill	November 1952	Edge Hill	June 1954	
45545 *Planet*	Camden	September 1956	Camden	February 1957	
45545 *Planet*	Willesden	May 1957	Upperby	June 1961	
5735 *Comet*	Camden	March 1945	Camden	April 1945	Loaned for Dynamometer trials between Crewe and Carlisle.
45736 *Phoenix*	Camden	September 1945	Holyhead	September 1955	
45736 *Phoenix*	Longsight	September 1960	Holyhead	September 1962	
46106 *Gordon Highlander*	Edge Hill	September 1952	Edge Hill	May 1953	
46106 *Gordon Highlander*	Edge Hill	September 1953	Camden	June 1957	
46106 *Gordon Highlander*	Newton Heath	September 1960	Trafford Park	December 1960	
46108 *Seaforth Highlander*	Longsight	September 1960	Upperby	November 1960	
46110 *Grenadier Guardsman*	Camden	September 1956	Kentish Town	October 1957	
46110 *Grenadier Guardsman*	Camden	September 1958	Edge Hill	October 1961	
46111 *Royal Fusilier*	Edge Hill	September 1952	Longsight	June 1953	
46111 *Royal Fusilier*	Longsight	September 1960	Willesden	June 1961	
46112 *Sherwood Forester*	Holyhead	February 1946	Holyhead	June 1946	
46112 *Sherwood Forester*	Holyhead	May 1947	Holyhead	June 1947	
46112 *Sherwood Forester*	Holyhead	August 1948	Holyhead	October 1948	

46116 *Irish Guardsman*	Camden	September 1958	Edge Hill	October 1961	
46118 *Royal Welch Fusilier*	Camden	September 1950	Holyhead	February 1953	On loan to Holyhead.
46118 *Royal Welch Fusilier*	Holyhead	April 1953	Camden	August 1954	
46118 *Royal Welch Fusilier*	Camden	September 1954	Nottingham	January 1960	
46119 *Lancashire Fusilier*	Camden	October 1946	Holbeck	December 1946	On loan to Leeds.
46119 *Lancashire Fusilier*	Holbeck	January 1947	Willesden	July 1947	On loan to Willesden.
46119 *Lancashire Fusilier*	Holyhead	August 1947	Longsight	November 1948	On loan to Longsight.
46119 *Lancashire Fusilier*	Longsight	November 1948	Holyhead	April 1950	
46119 *Lancashire Fusilier*	Holyhead	April 1953	Camden	April 1953	
46119 *Lancashire Fusilier*	Camden	June 1953	Holyhead	April 1954	
46119 *Lancashire Fusilier*	Camden	91954	Camden	June 1955	
46119 *Lancashire Fusilier*	Camden	September 1955	Camden	June 1958	
46119 *Lancashire Fusilier*	Camden	September 1958	Edge Hill	October 1958	
46120 *Royal Inniskilling Fusilier*	Longsight	October 1954	Derby	October 1955	Loaned to the CME dept for tests over the Peak District.
46120 *Royal Inniskilling Fusilier*	Longsight	June 1956	Camden	June 1958	
46120 *Royal Inniskilling Fusilier*	Camden	September 1958	Longsight	June 1960	
46120 *Royal Inniskilling Fusilier*	Llandudno Junction	June 1963	Withdrawn	July 1963	Withdrawn within two weeks of arriving at Crewe North. Stored at Crewe Works between July 1963 and September 1963 and scrapped there in October 1963.
46125 *3rd Carabinier*	Holyhead	August 1944	Holyhead	March 1945	
46125 *3rd Carabinier*	Holyhead	April 1945	Holyhead	June 1945	On loan.
46125 *3rd Carabinier*	Holyhead	July 1945	Holyhead	July 1947	On loan.
46125 *3rd Carabinier*	Holyhead	August 1947	Holbeck	October 1948	On loan.
46125 *3rd Carabinier*	Holbeck	February 1949	Holbeck	March 1949	On loan.
46125 *3rd Carabinier*	Holbeck	April 1949	Holbeck	May 1949	
46125 *3rd Carabinier*	Holbeck	August 1949	Camden	July 1952	
46125 *3rd Carabinier*	Camden	September 1952	Holyhead	February 1953	On loan.
46125 *3rd Carabinier*	Holyhead	March 1958	Camden	June 1953	
46125 *3rd Carabinier*	Camden	September 1953	Upperby	February 1954	
46125 *3rd Carabinier*	Upperby	March 1954	Upperby	May 1962	Only at Upperby for three weeks.
46125 *3rd Carabinier*	Upperby	May 1962	Holyhead	September 1962	
46125 *3rd Carabinier*	Holyhead	March 1963	Holyhead	April 1963	
46126 *Royal Army Service Corps*	Camden	September 1952	Camden	January 1953	
46127 *Old Contemptibles*	Holyhead	February 1951	Holyhead	March 1951	
46127 *Old Contemptibles*	Holyhead	October 1954	Camden	June 1955	

46127 *Old Contemptibles*	Camden	September 1955	Camden	July 1957	
46127 *Old Contemptibles*	Camden	September 1957	Kentish Town	October 1957	
46127 *Old Contemptibles*	Holyhead	April 1960	Longsight	June 1960	
46127 *Old Contemptibles*	Longsight	September 1960	Upperby	May 1962	
46128 *The Lovat Scouts*	Holbeck	December 1946	Upperby	May 1947	Initially on loan from Holbeck before the move was made permanent.
46128 *The Lovat Scouts*	Upperby	October 1948	Holyhead	June 1949	
46128 *The Lovat Scouts*	Holbeck	January 1950	Holyhead	April 1952	
46128 *The Lovat Scouts*	Holyhead	October 1952	Camden	June 1953	
46128 *The Lovat Scouts*	Camden	September 1953	Camden	June 1954	
46128 *The Lovat Scouts*	Camden	September 1954	Longsight	November 1954	
46128 *The Lovat Scouts*	Longsight	March 1955	Holyhead	May 1957	
46128 *The Lovat Scouts*	Holyhead	May 1957	Upperby	August 1962	
46128 *The Lovat Scouts*	Upperby	May 1962	Springs Branch	September 1962	
46129 *The Scottish Horse*	Holyhead	August 1954	Camden	June 1957	
46129 *The Scottish Horse*	Upperby	November 1957	Kingmoor	July 1961	
46129 *The Scottish Horse*	Upperby	March 1962	Longsight	September 1962	
46130 *The West Yorkshire Regiment*	Holyhead	July 1950	Holyhead	October 1950	On loan to Holyhead.
46130 *The West Yorkshire Regiment*	Holyhead	December 1950	Longsight	June 1953	
46131 *The Royal Warwickshire Regiment*	Holyhead	March 1945	Longsight	April 1947	
46131 *The Royal Warwickshire Regiment*	Trafford Park	July 1958	Camden	August 1958	
46131 *The Royal Warwickshire Regiment*	Camden	September 1958	Longsight	October 1958	
46132 *The King's Regiment Liverpool*	Holyhead	June 1945	Holbeck	September 1945	
46132 *The King's Regiment Liverpool*	Holbeck	October 1945	Holyhead	September 1946	
46132 *The King's Regiment Liverpool*	Holyhead	May 1952	Holyhead	July 1952	
46134 *The Cheshire Regiment*	Holyhead	February 1955	Holyhead	September 1958	
46134 *The Cheshire Regiment*	Holyhead	September 1958	Edge Hill	October 1961	
46135 *The East Lancashire*	Upperby	September 1956	Longsight	September 1956	
46135 *The East Lancashire*	Longsight	February 1957	Camden	June 1959	
46135 *The East Lancashire*	Camden	September 1959	Longsight	June 1960	
46135 *The East Lancashire*	Longsight	September 1960	Holbeck	August 1962	
46136 *The Border Regiment*	Upperby	October 1960	Holbeck	August 1962	

46137 *The Prince of Wales's Volunteers (South Lancashire)*	Longsight	February 1957	Holyhead	May 1957	
46138 *The London Irish Rifleman*	Camden	May 1952	Camden	April 1953	
46138 *The London Irish Rifleman*	Camden	June 1953	Camden	June 1953	
46138 *The London Irish Rifleman*	Camden	September 1953	Longsight	November 1954	
46138 *The London Irish Rifleman*	Longsight	January 1955	Upperby	August 1956	
46138 *The London Irish Rifleman*	Upperby	November 1956	Llandudno Junction	November 1959	
46139 *The Welch Regiment*	Camden	September 1945	Camden	December 1945	
46139 *The Welch Regiment*	Camden	October 1946	Holyhead	May 1947	
46143 *The South Staffordshire Regiment*	Longsight	February 1951	Camden	March 1951	
46143 *The South Staffordshire Regiment*	Holyhead	October 1952	Longsight	December 1952	
46144 *Honourable Artillery Company*	Edge Hill	February 1951	Edge Hill	March 1951	
46144 *Honourable Artillery Company*	Llandudno Junction	June 1963	Withdrawn	January 1964	Stored at Crewe North between September 1963 and January 1964, then withdrawn in January 1964 and scrapped at Crewe Works the same month.
46145 *The Duke of Wellington's Regiment (West Riding)*	Crewe Works	January 1944	Holyhead	February 1944	Initially on loan to Holyhead.
46145 *The Duke of Wellington's Regiment (West Riding)*	Holyhead	August 1946	Longsight	April 1947	
46146 *The Rifle Brigade*	Holbeck	December 1943	Holyhead	August 1944	
46146 *The Rifle Brigade*	Holyhead	May 1945	Holbeck	April 1946	
46146 *The Rifle Brigade*	Edge Hill	June 1947	Holyhead	April 1948	
46146 *The Rifle Brigade*	Holyhead	May 1948	Holyhead	December 1948	On loan to Holyhead.
46146 *The Rifle Brigade*	Holyhead	January 1949	Holyhead	May 1949	
46146 *The Rifle Brigade*	Holyhead	July 1949	Upperby	July 1951	
46148 *The Manchester Regiment*	Crewe Works	July 1954	Camden	June 1957	
46148 *The Manchester Regiment*	Camden	September 1957	Camden	September 1957	
46148 *The Manchester Regiment*	Willesden	April 1962	Llandudno Junction	September 1962	
46149 *The Middlesex Regiment*	Holbeck	September 1945	Camden	September 1945	
46149 *The Middlesex Regiment*	Holyhead	April 1960	Longsight	June 1961	
46150 *The Life Guardsman*	Upperby	January 1955	Edge Hill	April 1956	
46150 *The Life Guardsman*	Holyhead	September 1956	Upperby	September 1956	
46150 *The Life Guardsman*	Upperby	October 1956	Camden	June 1957	
46150 *The Life Guardsman*	Camden	September 1957	Llandudno Junction	November 1959	

46150 *The Life Guardsman*	Holyhead	March 1963	Holyhead	April 1963	
46151 *The Royal Horse Guardsman*	Crewe Works	April 1953	Holyhead	July 1955	
46151 *The Royal Horse Guardsman*	Camden	September 1956	Longsight	May 1958	
46151 *The Royal Horse Guardsman*	Longsight	June 1958	Longsight	July 1959	
46152 *The King's Dragoon Guardsman*	Crewe Works	August 1945	Camden	September 1945	
46152 *The King's Dragoon Guardsman*	Camden	September 1958	Holyhead	July 1960	
46152 *The King's Dragoon Guardsman*	Holyhead	August 1960	Willesden	June 1961	
46153 *The Royal Dragoon*	Edge Hill	February 1950	Edge Hill	March 1951	
45155 *The Lancer*	Holyhead	August 1951	Camden	July 1952	
45155 *The Lancer*	Camden	September 1952	Upperby	February 1957	Only at Upperby for two weeks.
45155 *The Lancer*	Upperby	February 1957	Edge Hill	June 1958	
45155 *The Lancer*	Edge Hill	September 1959	Camden	December 1960	Only at Camden for one week.
45155 *The Lancer*	Camden	December 1960	Llandudno Junction	September 1962	
45155 *The Lancer*	Llandudno Junction	June 1963	Holyhead	August 1963	
45155 *The Lancer*	Holyhead	September 1963	Kingmoor	September 1964	
46156 *The South Wales Borderer*	Crewe Works	May 1954	Upperby	August 1956	
46156 *The South Wales Borderer*	Upperby	September 1956	Edge Hill	June 1958	
46157 *The Royal Artilleryman*	Crewe Works	January 1946	Holyhead	February 1946	On loan.
46157 *The Royal Artilleryman*	Holyhead	May 1946	Holyhead	12 1947	
46157 *The Royal Artilleryman*	Holyhead	December 1947	Holyhead	June 1948	
46157 *The Royal Artilleryman*	Holyhead	October 1948	Holyhead	January 1949	On loan to Holyhead
46157 *The Royal Artilleryman*	Holyhead	February 1949	Longsight	February 1949	On loan to Longsight.
46157 *The Royal Artilleryman*	Edge Hill	July 1949	Holyhead	March 1951	
46157 *The Royal Artilleryman*	Camden	September 1958	Camden	June 1959	
46159 *The Royal Air Force*	Holyhead	November 1953	Camden	June 1954	
46159 *The Royal Air Force*	Camden	September 1954	Camden	June 1955	
46159 *The Royal Air Force*	Camden	September 1955	Camden	November 1956	
46159 *The Royal Air Force*	Camden	December 1956	Willesden	January 1961	
46160 *Queen Victoria's Rifleman*	Crewe Works	February 1945	Holbeck	June 1945	On loan to Leeds for two weeks.
46160 *Queen Victoria's Rifleman*	Holbeck	June 1945	Holyhead	August 1946	
46161 *King's Own*	Crewe Works	October 1946	Holyhead	April 1950	
46161 *King's Own*	Holyhead	October 1951	Camden	July 1952	

46161 *King's Own*	Camden	September 1952	Longsight	June 1953	
46161 *King's Own*	Camden	September 1956	Camden	June 1959	
46161 *King's Own*	Springs Branch	July 1962	Holbeck	July 1962	
46162 *Queen's Westminster Rifleman*	Crewe Works	January 1948	Holyhead	March 1948	
46162 *Queen's Westminster Rifleman*	Holyhead	April 1948	Camden	April 1948	
46162 *Queen's Westminster Rifleman*	Camden	September 1952	Camden	October 1952	
46163 *Civil Service Rifleman*	Camden	September 1954	Camden	June 1959	
46163 *Civil Service Rifleman*	Preston	April 1960	Holyhead	June 1960	
46164 *The Artists' Rifleman*	Edge Hill	September 1959	Millhouses	February 1960	
46165 *The Ranger 12th London Regiment*	Upperby	August 1956	Upperby	November 1956	
46165 *The Ranger 12th London Regiment*	Upperby	February 1957	Upperby	February 1957	
46165 *The Ranger 12th London Regiment*	Springs Branch	July 1962	Llandudno Junction	September 1962	
46165 *The Ranger 12th London Regiment*	Llandudno Junction	June 1963	Annesley	February 1964	
46166 *London Rifle Brigade*	Crewe Works	January 1945	Edge Hill	May 1946	On loan to Edge Hill.
46166 *London Rifle Brigade*	Edge Hill	May 1946	Holyhead	September 1946	
46166 *London Rifle Brigade*	Holyhead	October 1946	Holyhead		
46166 *London Rifle Brigade*	Holyhead	December 1948	Holyhead	May 1949	
46166 *London Rifle Brigade*	Holyhead	October 1949	Longsight	January 1950	
46166 *London Rifle Brigade*	Longsight	February 1950	Holyhead	June 1950	
46166 *London Rifle Brigade*	Holyhead	February 1951	Holyhead	March 1951	
46166 *London Rifle Brigade*	Holyhead	March 1951	Edge Hill	April 1951	
46166 *London Rifle Brigade*	Edge Hill	June 1951	Camden	July 1952	
46166 *London Rifle Brigade*	Camden	September 1952	Edge Hill	June 1953	
46166 *London Rifle Brigade*	Edge Hill	June 1953	Camden	June 1953	
46166 *London Rifle Brigade*	Camden	September 1954	Longsight	November 1954	
46166 *London Rifle Brigade*	Longsight	January 1955	Holyhead	October 1955	
46166 *London Rifle Brigade*	Upperby	December 1955	Longsight	June 1958	
46166 *London Rifle Brigade*	Longsight	September 1960	Kingmoor	June 1961	
46166 *London Rifle Brigade*	Kingmoor	July 1961	Upperby	May 1962	
46167 *The Hertfordshire Regiment*	Crewe Works	December 1948	Longsight	October 1949	
46167 *The Hertfordshire Regiment*	Longsight	December 1951	Camden	March 1953	
46167 *The Hertfordshire Regiment*	Camden	September 1954	Upperby	October 1956	
46167 *The Hertfordshire*	Springs	June 1962	Holyhead	September	

Regiment	Branch			1962	
46167 *The Hertfordshire Regiment*	Holyhead	March 1963	Holyhead	April 1963	
46168 *The Girl Guide*	Crewe Works	April 1946	Holyhead	August 1946	
46168 *The Girl Guide*	Holyhead	September 1946	Camden	November 1946	
46169 *The Boy Scout*	Longsight	July 1959	Willesden	April 1962	
46170 *British Legion*	Crewe Works	November 1935	Longsight	November 1935	
46170 *British Legion*	Longsight	February 1937	Longsight	November 1938	
46170 *British Legion*	Longsight	July 1939	Holbeck	July 1943	On loan to Holbeck.
46170 *British Legion*	Camden	September 1958	Camden	October 1958	
46170 *British Legion*	Camden	November 1959	Preston	December 1959	
46170 *British Legion*	Preston	January 1960	Holyhead	June 1960	
46170 *British Legion*	Holyhead	July 1960	Camden	August 1960	
46170 *British Legion*	Camden	September 1960	Camden	October 1960	
46170 *British Legion*	Camden	December 1960	Llandudno Junction	September 1962	

Appendix 4: Holyhead Allocation

Loco	From	Date	To	Date	Comments
45514 *Holyhead*	Crewe Works	May 1947	Crewe North	June 1947	
45514 *Holyhead*	Bushbury	August 1948	Bushbury	October 1948	
45521 *Rhyl*	Longsight	May 1947	Edge Hill	June 1947	
45527 *Southport*	Bushbury	March 1961	Llandudno Junction	June 1961	
45527 *Southport*	Llandudno Junction	September 1961	Willesden	June 1963	Stored serviceable early 1963.
45528 *REME*	Longsight	March 1955	Crewe North	May 1955	
45528 *REME*	Crewe North	April 1957	Crewe North	April 1957	
45530 *Sir Frank Ree*	Willesden	June 1964	Kingmoor	January 1965	Stored 1 November 1964 to December 1964.
45532 *Illustrious*	Crewe Works	July 1948	Bushbury	August 1948	On loan.
45532 *Illustrious*	Camden	July 1950	Camden	August 1950	On loan.
45534 *E. Tootal Broadhurst*	Longsight	July 1950	Longsight	October 1950	On loan.
45736 *Phoenix*	Crewe North	September 1955	Longsight	June 1960	
45736 *Phoenix*	Crewe North	September 1962	Camden	March 1963	
46111 *Royal Fusilier*	Longsight	August 1953	Longsight	September 1953	
46111 *Royal Fusilier*	Longsight	May 1957	Longsight	June 1957	
46112 *Sherwood Forester*	Crewe North	December 1943	Crewe North	February 1946	
46112 *Sherwood Forester*	Crewe North	April 1946	Crewe North	May 1947	

46112 *Sherwood Forester*	Crewe North	June 1947	Crewe North	August 1948	
46112 *Sherwood Forester*	Crewe North	October 1948	Holbeck	February 1953	
46114 *Coldstream Guardsman*	Llandudno Junction	September 1961	Willesden	June 1963	Stored serviceable in early 1963.
46116 *Irish Guardsman*	Upperby	April 1956	Upperby	June 1956	
46118 *Royal Welch Fusilier*	Edge Hill	May 1947	Edge Hill	June 1947	
46118 *Royal Welch Fusilier*	Edge Hill	August 1947	Edge Hill	September 1947	
46118 *Royal Welch Fusilier*	Crewe North	February 1953	Crewe North	April 1953	On loan.
46119 *Lancashire Fusilier*	Willesden	August 1947	Crewe North	August 1947	On loan.
46119 *Lancashire Fusilier*	Crewe North	April 1950	Crewe North	April 1953	
46119 *Lancashire Fusilier*	Crewe North	April 1954	Camden	June 1954	
46120 *Royal Inniskilling Fusilier*	Longsight	June 1954	Longsight	July 1954	
46123 *Royal Irish Fusilier*	Edge Hill	May 1958	Edge Hill	June 1958	
46125 *3rd Carabinier*	Crewe North	March 1945	Crewe North	April 1945	
46125 *3rd Carabinier*	Crewe North	June 1945	Crewe North	July 1945	On loan.
46125 *3rd Carabinier*	Crewe North	July 1947	Crewe North	August 1947	
46125 *3rd Carabinier*	Crewe North	February 1953	Crewe North	March 1958	
46125 *3rd Carabinier*	Crewe North	September 1962	Crewe North	March 1963	
46125 *3rd Carabinier*	Crewe North	April 1963	Willesden	June 1963	
46126 *Royal Army Service Corps*	Crewe North	May 1946	Crewe North	July 1947	On loan.
46126 *Royal Army Service Corps*	Carlisle	January 1958	Upperby	February 1958	
46127 *Old Contemptibles*	Crewe North	August 1944	Crewe North	June 1946	On loan.
46127 *Old Contemptibles*	Edge Hill	June 1947	Crewe North	February 1951	
46127 *Old Contemptibles*	Crewe North	March 1951	Crewe North	October 1954	
46127 *Old Contemptibles*	Kentish Town	July 1958	Crewe North	April 1960	
46128 *The Lovat Scouts*	Crewe North	June 1949	Holbeck	November 1949	On loan.
46128 *The Lovat Scouts*	Crewe North	May 1952	Crewe North	October 1952	
46128 *The Lovat Scouts*	Crewe North	May 1957	Crewe North	May 1957	
46129 *The Scottish Horse*	Longsight	February 1953	Crewe North	August 1954	
46130 *The West Yorkshire Regiment*	Crewe North	June 1950	Crewe North	July 1950	
46130 *The West Yorkshire Regiment*	Crewe North	October 1950	Crewe North	December 1950	
46130 *The West Yorkshire Regiment*	Edge Hill	July 1957	Edge Hill	October 1957	
46131 *The Royal Warwickshire Regiment*	Crewe North	February 1945	Crewe North	March 1945	

46132 *The King's Regiment Liverpool*	Crewe North	June 1945	Crewe North	June 1945	
46132 *The King's Regiment Liverpool*	Crewe North	September 1946	Edge Hill	May 1947	
46132 *The King's Regiment Liverpool*	Edge Hill	June 1947	Crewe North	May 1952	
46132 *The King's Regiment Liverpool*		July 1952	Edge Hill	October 1954	
46134 *The Cheshire Regiment*	Camden	November 1954	Crewe North	February 1955	
46134 *The Cheshire Regiment*	Crewe North	September 1958	Crewe North	September 1958	
46135 *The East Lancashire Regiment*	Edge Hill	April 1954	Carlisle	July 1954	
46137 *The Prince of Wales's Volunteers (South Lancashire)*	Crewe North	May 1957	Upperby	August 1957	
46138 *The London Irish Rifleman*	Edge Hill	September 1951	Camden	April 1952	
46138 *The London Irish Rifleman*	Llandudno Junction	June 1960	Llandudno Junction	September 1960	
46138 *The London Irish Rifleman*	Llandudno Junction	March 1961	Upperby	November 1962	
46139 *The Welch Regiment*	Crewe North	April 1947	Camden	June 1947	
46143 The South Staffordshire Regiment	Camden	September 1952	Crewe North	October 1952	
46145 *The Duke of Wellington's Regiment (West Riding)*	Crewe North	February 1944	Crewe North	August 1944	On loan.
46145 *The Duke of Wellington's Regiment (West Riding)*	Crewe North	August 1944	Crewe North	August 1946	
46145 *The Duke of Wellington's Regiment (West Riding)*	Longsight	July 1951	Longsight	September 1951	On loan.
46145 *The Duke of Wellington's Regiment (West Riding)*	Longsight	October 1951	Longsight	November 1951	On loan.
46146 *The Rifle Brigade*	Crewe North	April 1948	Crewe North	May 1948	
46146 *The Rifle Brigade*	Crewe North	December 1948	Crewe North	January 1949	
46146 *The Rifle Brigade*	Crewe North	May 1949	Crewe North	July 1949	
46147 *The Northamptonshire Regiment*	Camden	September 1955	Edge Hill	June 1956	
46147 *The Northamptonshire Regiment*	Longsight	September 1956	Edge Hill	June 1957	The EHC has the engine transferred to Holyhead and Longsight at the same time.
46147 *The Northamptonshire Regiment*	Edge Hill	September 1957	Upperby	October 1957	
46147 *The Northamptonshire Regiment*	Upperby	November 1957	Edge Hill	June 1958	
46148 *The Manchester Regiment*	Llandudno Junction	December 1963	Llandudno Junction	January 1964	
46148 *The Manchester Regiment*	Llandudno Junction	April 1964	Withdrawn	November 1964	Stored at Birkenhead November 1964 to December 1964. Scrapped at Birds, Morriston, Swansea, January 1965.

46149 *The Middlesex Regiment*	Longsight	December 1949	Longsight	January 1949	On loan.
46149 *The Middlesex Regiment*	Longsight	September 1956	Crewe North	April 1960	
46150 *The Life Guardsman*	Longsight	July 1951	Upperby	August 1954	
46150 *The Life Guardsman*	Edge Hill	August 1956	Crewe North	September 1956	
46150 *The Life Guardsman*	Llandudno Junction	June 1960	Crewe North	March 1963	
46150 *The Life Guardsman*	Crewe North	April 1963	Willesden	June 1963	
46151 *The Royal Horse Guardsman*	Crewe North	July 1955	Camden	June 1956	
46152 *The King's Dragoon Guardsman*	Crewe North	July 1960	Crewe North	August 1960	
46152 *The King's Dragoon Guardsman*	Llandudno Junction	March 1963	Kingmoor	January 1965	
46154 *The Hussar*	Llandudno Junction	March 1962	Willesden	June 1962	
45155 *The Lancer*	Crewe North	August 1963	Kingmoor	September 1964	
46156 *The South Wales Borderer*	Llandudno Junction	June 1960	Camden	March 1963	
46156 *The South Wales Borderer*	Camden	April 1963	Willesden	June 1963	
46156 *The Royal Artilleryman*	Crewe North	February 1946	Crewe North	May 1946	On loan.
46156 *The Royal Artilleryman*	Crewe North	December 1946	Crewe North	January 1947	
46156 *The Royal Artilleryman*	Crewe North	June 1948	Crewe North	October 1948	
46156 *The Royal Artilleryman*	Crewe North	January 1949	Crewe North	February 1949	
46156 *The Royal Artilleryman*	Crewe North	March 1951	Camden	August 1954	
46159 *The Royal Air Force*	Camden	July 1953	Crewe North	November 1953	
46160 *Queen Victoria's Rifleman*	Crewe North	August 1946	Longsight	May 1947	
46161 *King's Own*	Crewe North	April 1950	Crewe North	October 1951	
46161 *King's Own*	Longsight	July 1953	Longsight	October 1953	
46162 *Queen's Westminster Rifleman*	Crewe North	March 1948	Crewe North	April 1948	
46163 *Civil Service Rifleman*	Preston	April 1960	Preston	April 1960	
46163 *Civil Service Rifleman*	Crewe North	June 1960	Llandudno Junction	September 1960	
46163 *Civil Service Rifleman*	Llandudno Junction	March 1961	Willesden	June 1962	
46166 *London Rifle Brigade*	Crewe North	September 1946	Crewe North	October 1946	On loan.
46166 *London Rifle Brigade*	Crewe North	May 1949	Crewe North	October 1949	
46166 *London Rifle Brigade*	Crewe North	June 1950	Crewe North	February 1951	Also loaned for a few days in March 1951.
46166 *London Rifle Brigade*	Crewe North	October 1955	Upperby	November 1955	
46167 *The Hertfordshire Regiment*	Crewe North	September 1962	Crewe North	March 1963	

46167 *The Hertfordshire Regiment*	Crewe North	April 1963	Willesden	June 1963	

Appendix 5: Llandudno Junction Allocation

Loco	From	Date	To	Date	Comments
45525 *Colwyn Bay*	Willesden	September 1961	Withdrawn	May 1963	Stored serviceable between 4 November 1962 and 3 February 1963. Stored at Crewe Works, May 1963 and scrapped there in June 1963.
45527 *Southport*	Holyhead	June 1961	Holyhead	September 1961	
45530 *Sir Frank Ree*	Willesden	September 1961	Willesden	July 1962	
45534 *E. Tootal Broadhurst*	Edge Hill	November 1959	Longsight	June 1960	
45534 *E. Tootal Broadhurst*	Willesden	September 1961	Crewe North	June 1963	Stored 4 November 1962 to 3 February 1963.
46101 *Royal Scots Grey*	Willesden	September 1961	Willesden	June 1962	
46114 *Coldstream Guardsman*	Bushbury	June 1961	Holyhead	September 1961	
46120 *Royal Inniskilling Fusilier*	Willesden	September 1961	Crewe North	June 1963	
46138 *The London Irish Rifleman*	Crewe North	November 1959	Holyhead	June 1960	
46138 *The London Irish Rifleman*	Holyhead	September 1960	Holyhead	March 1961	
46144 *Honourable Artillery Company*	Willesden	September 1961	Crewe North	June 1963	
46148 *The Manchester Regiment*	Crewe North	September 1962	Holyhead	December 1963	Seen ex-Works in April 1963.
46148 *The Manchester Regiment*	Holyhead	January 1964	Holyhead	April 1964	Stored serviceable in November 1963
46150 *The Life Guard*	Crewe North	November 1959	Holyhead	June 1960	
46152 *The King's Dragoon Guardsman*	Willesden	September 1961	Holyhead	March 1962	
46154 *The Hussar*	Willesden	September 1961	Holyhead	March 1962	
46155 *The Lancer*	Crewe North	September 1962	Crewe North	June 1963	
46156 *The South Wales Borderer*	Edge Hill	November 1959	Longsight	February 1960	
46156 *The South Wales Borderer*	Longsight	March 1960	Holyhead	June 1960	
46163 *Civil Service Rifleman*	Holyhead	September 1960	Holyhead	March 1961	
46165 *The Ranger 12th London Regiment*	Crewe North	September 1962	Crewe North	June 1963	Seen ex-Works at Chester in November 1962.
46170 *British Legion*	Crewe North	September 1962	Withdrawn	November 1962	Stored September 1962 to November 1962 at Llandudno. Loco had been stored in anticipation of a return to traffic, as motion greased and chimney covered. Stored Crewe Works, from November 1962 and scrapped in January 1963.

Appendix 6: Edge Hill Allocation

Loco	From	Date	To	Date	Comments
45521 *Rhyl*	Holyhead	June 1947	Springs Branch	September 1961	
45525 *Colwyn Bay*	Crewe North	October 1952	Willesden	January 1961	Initially on loan.
45526 *Morecambe and Heysham*	Holbeck	July 1947	Bushbury	October 1949	
45527 *Southport*	Crewe Works	September 1948	Bushbury	March 1961	
45531 *Sir Frederick Harrison*	Bushbury	June 1950	Springs Branch	October 1963	
45534 *E. Tootal Broadhurst*	Camden	November 1951	Llandudno Junction	November 1959	
45535 *Sir Herbert Walke, K.C.B.*	Camden	November 1950	Crewe North	December 1950	On loan.
45535 *Sir Herbert Walker, K.C.B.*	Crewe North	October 1952	Crewe North	November 1952	On loan.
45535 *Sir Herbert Walker, K.C.B.*	Crewe North	June 1954	Kingmoor	November 1962	
46106 *Gordon Highlander*	Crewe North	May 1953	Crewe North	September 1953	
46110 *Grenadier Guardsman*	Holyhead	March 1954	Camden	June 1956	
46110 *Grenadier Guardsman*	Crewe North	October 1961	Longsight	October 1961	
46110 *Grenadier Guardsman*	Longsight	December 1961	Springs Branch	June 1963	
46114 *Coldstream Guardsman*	Camden	September 1955	Longsight	February 1956	
46114 *Coldstream Guardsman*	Longsight	February 1956	Willesden	January 1961	
46116 *Irish Guardsman*	Crewe North	October 1961	Kingmoor	November 1962	
46118 *Royal Welch Fusilier*	Crewe Works	December 1946	Holyhead	May 1947	
46118 *Royal Welch Fusilier*	Holyhead	June 1947	Holyhead	August 1947	
46118 *Royal Welch Fusilier*	Holyhead	September 1947	Camden	October 1947	
46119 *Lancashire Fusilier*	Crewe North	October 1958	Withdrawn	November 1963	
46123 *Royal Irish Fusilier*	Crewe Works	May 1949	Holyhead	May 1958	
46123 *Royal Irish Fusilier*	Holyhead	June 1958	Kentish Town	September 1959	
46124 *London Scottish*	Crewe North	March 1945	Camden	October 1947	
46124 *London Scottish*	Camden	December 1947	Kingmoor	November 1962	
46126 *Royal Army Service Corps*	Camden	December 1948	Holbeck	September 1949	
46126 *Royal Army Service Corps*	Preston	September 1960	Willesden	January 1961	
46127 *Old Contemptibles*	Crewe North	May 1947	Holyhead	June 1947	
46130 *The West Yorkshire Regiment*	Carlisle	June 1957	Holyhead	July 1957	
46130 *The West Yorkshire*	Holyhead	October	Kentish Town	February	

		1957		1959	
Regiment					
46130 *The West Yorkshire Regiment*	Kentish Town	April 1959	Kentish Town	September 1959	
46132 *The King's Regiment Liverpool*	Holyhead	May 1947	Holyhead	June 1947	
46132 *The King's Regiment Liverpool*	Holyhead	October 1954	Longsight	November 1954	
46132 *The King's Regiment Liverpool*	Longsight	January 1955	Kentish Town	September 1959	
46134 *The Cheshire Regiment*	Crewe North	October 1961	Upperby	May 1962	
46135 *The East Lancashire Regiment*	Crewe Works	January 1947	Holyhead	April 1954	
46137 *The Prince of Wales's Volunteers (South Lancashire)*	Longsight	October 1957	Longsight	December 1957	
46142 *The York and Lancaster Regiment*	Camden	September 1955	Kentish Town	September 1959	
46114 *Honourable Artillery Company*	Crewe North	March 1951	Camden	December 1952	
46146 *The Rifle Brigade*	Crewe North	April 1947	Crewe North	June 1947	
46147 *The Northamptonshire Regiment*	Crewe North	June 1948	Upperby	March 1949	
46147 *The Northamptonshire Regiment*	Holyhead	June 1956	Longsight	June 1956	
46147 *The Northamptonshire Regiment*	Holyhead	June 1958	Crewe North	September 1959	
46149 *The Middlesex Regiment*	Longsight	January 1953	Longsight	June 1956	
46150 *The Life Guardsman*	Crewe North	April 1956	Holyhead	August 1956	
46152 *King's Dragoon Guardsman*	Longsight	January 1953	Kentish Town	October 1957	
46153 *The Royal Dragoon*	Crewe North	March 1951	Longsight	July 1955	
46154 *The Hussar*	Preston	September 1960	Willesden	July 1961	
46155 *The Lancer*	Crewe North	June 1958	Crewe North	September 1959	
46156 *The South Wales Borderer*	Crewe North	June 1958	Llandudno Junction	November 1959	
46157 *The Royal Artilleryman*	Longsight	March 1949	Crewe North	July 1949	
46157 *The Royal Artilleryman*	Longsight	October 1954	Kentish Town	October 1957	
46164 *The Artists' Rifleman*	Crewe Works	June 1951	Crewe North	September 1959	
46166 *London Rifle Brigade*	Crewe North	May 1946	Crewe North	May 1946	On loan for two weeks.
46166 *London Rifle Brigade*	Crewe North	April 1951	Crewe North	June 1951	On loan.
46166 *London Rifle Brigade*	Crewe North	June 1953	Crewe North	September 1953	

Appendix 7: Camden Allocation

Loco	From	Date	To	Date	Comments
45514 *Holyhead*	Bushbury	June 1950	Sheffield Millhouses	February 1960	
45522 *Prestatyn*	Bushbury	June 1950	Kentish Town	November 1959	
45523 *Bangor*	Crewe North	July 1951	Willesden	January 1961	
45525 *Colwyn Bay*	Crewe North	July 1951	Crewe North	September 1951	
45525 *Colwyn Bay*	Crewe North	July 1952	Crewe North	September 1952	
45528 *REME*	Crewe North	July 1951	Crewe North	September 1951	
45528 *REME*	Crewe North	June 1953	Crewe North	September 1953	
45528 *REME*	Crewe North	June 1954	Crewe North	September 1954	
45529 *Stephenson*	Crewe North	July 1952	Crewe North	September 1952	
45529 *Stephenson*	Crewe North	June 1953	Crewe North	September 1953	
45529 *Stephenson*	Crewe North	November 1953	Crewe North	January 1954	
45529 *Stephenson*	Crewe North	March 1958	Crewe North	April 1958	
45530 *Sir Frank Ree*	Longsight	September 1960	Willesden	June 1961	
45531 *Sir Frederick Harrison*	Bushbury	May 1949	Bushbury	October 1949	
45532 *Illustrious*	Bushbury	May 1949	Holyhead	July 1950	
45532 *Illustrious*	Holyhead	August 1950	Nottingham	November 1959	
45534 *E. Tootal Broadhurst*	Longsight	August 1951	Edge Hill	November 1951	
45535 *Sir Herbert Walker, K.C.B.*	Crewe North	July 1951	Crewe North	September 1951	
45545 *Planet*	Bushbury	May 1949	Crewe North	September 1956	
45545 *Planet*	Crewe North	February 1957	Crewe North	May 1957	
45735 *Comet*	Holbeck	July 1943	Bushbury	October 1948	On loan to Crewe for three weeks in March 1945.
45735 *Comet*	Bushbury	November 1948	Preston	September 1959	
45736 *Phoenix*	Holbeck	July 1943	Crewe North	September 1954	
45736 *Phoenix*	Holyhead	March 1963	Holyhead	April 1963	
46101 *Royal Scots Grey*	Longsight	February 1949	Crewe North	September 1950	
46101 *Royal Scots Grey*	Crewe North	July 1952	Crewe North	September 1952	
46101 *Royal Scots Grey*	Crewe North	June 1953	Crewe North	September 1953	
46101 *Royal Scots Grey*	Crewe North	August 1954	Crewe North	September 1954	
46101 *Royal Scots Grey*	Crewe North	October 1954	Crewe North	January 1955	
46101 *Royal Scots Grey*	Crewe North	June 1959	Willesden	June 1961	
46106 *Gordon Highlander*	Crewe North	June 1957	Upperby	September 1957	
6109 *Royal Engineer*	Edge Hill	August 1943	Holbeck	August 1943	

46110 *Grenadier Guardsman*	Edge Hill	June 1956	Crewe North	September 1956	
46110 *Grenadier Guardsman*	Kentish Town	June 1958	Crewe North	September 1958	
46114 *Coldstream Guardsman*	Longsight	June 1955	Edge Hill	September 1955	
46116 *Irish Guardsman*	Trafford Park	July 1958	Crewe North	September 1958	
46118 *Royal Welch Fusilier*	Edge Hill	October 1947	Crewe North	September 1950	
46118 *Royal Welch Fusilier*	Crewe North	August 1954	Crewe North	September 1954	
46119 *Lancashire Fusilier*	Crewe Works	September 1944	Crewe North	October 1946	
46119 *Lancashire Fusilier*	Crewe North	April 1953	Crewe North	June 1953	
46119 *Lancashire Fusilier*	Holyhead	June 1954	Crewe North	September 1954	
46119 *Lancashire Fusilier*	Crewe North	June 1955	Crewe North	September 1955	
46119 *Lancashire Fusilier*	Crewe North	June 1958	Crewe North	September 1958	
46120 *Royal Inniskilling Fusilier*	Crewe North	June 1958	Crewe North	September 1958	
46120 *Royal Inniskilling Fusilier*	Crewe North	September 1960	Willesden	January 1961	
46121 *Highland Light Infantry, City of Glasgow Regiment*	Crewe Works	August 1946	Longsight	October 1946	
46122 *Royal Ulster Rifleman*	Crewe Works	September 1945	Longsight	October 1946	
46124 *London Scottish*	Edge Hill	October 1947	Edge Hill	December 1947	On loan.
46125 *3rd Carabinier*	Crewe North	July 1952	Crewe North	September 1952	
46125 *3rd Carabinier*	Crewe North	June 1953	Crewe North	September 1953	
46126 *The Royal Army Service Corps*	Crewe North	July 1948	Edge Hill	December 1948	
46126 *The Royal Army Service Corps*	Holbeck	101949	Crewe North	September 1952	
46126 *The Royal Army Service Corps*	Crewe North	January 1953	Upperby	September 1954	
46127 *Old Contemptibles*	Crewe North	June 1955	Crewe North	September 1955	
46127 *Old Contemptibles*	Crewe North	July 1957	Crewe North	September 1957	
46128 *The Lovat Scouts*	Crewe North	June 1953	Crewe North	September 1953	
46128 *The Lovat Scouts*	Crewe North	June 1954	Crewe North	September 1954	
46129 *The Scottish Horse*	Crewe North	June 1957	Crewe North	September 1957	
46129 *The Scottish Horse*	Crewe North				
46131 *The Royal Warwickshire Regiment*	Longsight	October 1947	Longsight	November 1947	On loan.
46131 *The Royal Warwickshire Regiment*	Crewe North	August 1958	Crewe North	September 1958	
46134 *The Cheshire Regiment*	Crewe North	October 1954	Holyhead	November 1954	
46135 *The East Lancashire Regiment*	Crewe North	June 1959	Crewe North	September 1959	

46138 *The London Irish Rifleman*	Holyhead	April 1952	Crewe North	May 1952	
46138 *The London Irish Rifleman*	Crewe North	April 1953	Crewe North	June 1953	
46138 *The London Irish Rifleman*	Crewe North	June 1953	Crewe North	September 1953	
46139 *The Welch Regiment*	Holyhead	June 1947	Kentish Town	November 1959	
46141 *The North Staffordshire Regiment*	Longsight	March 1951	Carlisle	September 1954	
46143 *The South Staffordshire Regiment*	Crewe North	February 1951	Holyhead	September 1952	
46144 *Honourable Artillery Company*	Edge Hill	December 1952	Willesden	June 1961	
46145 *The Duke of Wellington's Regiment (West Riding)*	Longsight	April 1947	Holyhead	December 1949	
46146 *The Rifle Brigade*	Upperby	April 1952	Upperby	January 1954	
46146 *The Rifle Brigade*	Upperby	March 1954	Willesden	January 1961	
46147 *The Northamptonshire Regiment*	Upperby	April 1952	Holyhead	September 1955	
46148 *The Manchester Regiment*	Crewe North	June 1957	Crewe North	September 1957	
46148 *The Manchester Regiment*	Crewe North	September 1957	Upperby	October 1957	
46149 *The Middlesex Regiment*	Crewe North	September 1945	Longsight	October 1946	
46150 *The Life Guardsman*	Crewe North	June 1957	Crewe North	September 1957	
46151 *The Royal Horse Guardsman*	Holyhead	June 1956	Crewe North	September 1956	
46152 *The King's Dragoon Guardsman*	Crewe Works	September 1945	Longsight	February 1951	
46152 *The King's Dragoon Guardsman*	Trafford Park	July 1958	Crewe North	September 1958	
46154 *The Hussar*	Crewe Works	April 1948	Kentish Town	May 1959	
46154 *The Hussar*	Kentish Town	June 1959	Preston	September 1959	
46155 *The Lancer*	Crewe North	July 1952	Crewe North	September 1952	
46155 *The Lancer*	Crewe North	December 1960	Crewe North	December 1960	One week.
46156 *The South Wales Borderer*	Holyhead	March 1963	Holyhead	April 1963	
46157 *The Royal Artilleryman*	Holyhead	August 1954	Longsight	September 1954	
46159 *The Royal Air Force*	Crewe Works	September 1945	Holyhead	July 1953	
46159 *The Royal Air Force*	Crewe North	June 1954	Crewe North	September 1954	
46159 *The Royal Air Force*	Crewe North	June 1955	Crewe North	September 1955	
46159 *The Royal Air Force*	Crewe North	November 1956	Crewe North	December 1956	
46161 *King's Own*	Crewe North	July 1952	Crewe North	September 1952	
46161 *King's Own*	Longsight	June 1956	Crewe North	September 1956	
46161 *King's Own*	Crewe North	June 1959	Preston	September 1959	

46162 *Queen's Westminster Rifleman*	Crewe North	April 1948	Longsight	January 1949	
46162 *Queen's Westminster Rifleman*	Longsight	April 1949	Crewe North	September 1952	
46162 *Queen's Westminster Rifleman*	Crewe North	October 1952	Kentish Town	November 1959	Initially on loan.
46163 *Civil Service Rifleman*	Crewe Works	October 1953	Crewe North	September 1954	
46163 *Civil Service Rifleman*	Crewe North	June 1959	Preston	September 1959	
46166 *London Rifle Brigade*	Camden	July 1952	Crewe North	September 1952	
46166 *London Rifle Brigade*	Crewe North	June 1953	Crewe North	September 1954	
46167 *The Hertfordshire Regiment*	Crewe North	March 1953	Crewe North	September 1953	
46167 *The Hertfordshire Regiment*	Crewe North	June 1954	Crewe North	September 1954	
46168 *The Girl Guide*	Crewe North	November 1946	Preston	September 1959	
46170 *British Legion*	Holbeck	August 1943	Edge Hill	September 1945	
46170 *British Legion*	Edge Hill	October 1945	Crewe North	September 1958	
46170 *British Legion*	Crewe North	October 1958	Crewe North	November 1959	
46170 *British Legion*	Crewe North	August 1960	Crewe North	September 1960	
46170 *British Legion*	Crewe North	October 1960	Crewe North	1 February 1960	

Appendix 8: Willesden Allocation

Loco	From	Date	To	Date	Comments
45523 *Bangor*	Camden	January 1961	Withdrawal	February 1964	Stored from December 1963 to February 1964 and scrapped at Crewe Works, March 1964.
45525 *Colwyn Bay*	Edge Hill	January 1961	Llandudno Junction	September 1961	
45527 *Southport*	Holyhead	June 1963	Kingmoor	September 1963	
45528 *REME*	Crewe North	January 1961	Withdrawal	January 1963	Stored at Willesden, September 1962 until withdrawal and then stored at Crewe Works, January 1963 to February 1963 and scrapped March 1963.
45529 *Stephenson*	Crewe North	January 1961	Annesley	October 1963	Stored at Willesden, December 1963 to February 1964 and withdrawn in February 1964. Scrapped Crewe Works, March 1964.
45735 *Comet*	Edge Hill	January 1961	Annesley	October 1963	
45736 *Phoenix*	Holyhead	June 1963	Upperby	July 1964	
46101 *Royal Scots Grey*	Camden	June 1960	Annesley	February 1963	Stored at Willesden, August 1963 to March 1964, having suffered a damaged cylinder whilst on shed at Willesden and then withdrawn on September 1963.

46111 *Royal Fusilier*	Crewe North	June 1961	Annesley	February 1963	
46114 *Coldstream Guardsman*	Holyhead	June 1963	Annesley	September 1963	Some question whether it reached Annesley, but Annesley staff said that it arrived but they withdrew it immediately. Stored at Willesden, November 1963 to March 1964 before moving to Rotherham and scrapping April 1964.
46120 *Royal Inniskilling Fusilier*	Camden	January 1961	Llandudno Junction	September 1961	
46122 *Royal Ulster Rifleman*	Bushbury	December 1960	Trafford Park	December 1961	
46125 *3rd Carabinier*	Holyhead	June 1963	Annesley	September 1963	
46126 *Royal Army Service Corps*	Edge Hill	January 1961	Annesley	December 1962	
46141 *The North Staffordshire Regiment*	Bushbury	December 1960	Trafford Park	December 1960	Only at Willesden for three weeks.
46144 *Honourable Artillery*	Camden	June 1961	Llandudno Junction	September 1961	
46146 *The Rifle Brigade*	Camden	January 1961	Withdrawn	November 1962	Stored at Devons Road Bow, November 1962 to March 1963 and scrapped at Crewe Works, March 1963.
46150 *The Life Guardsman*	Holyhead	June 1963	Withdrawal	December 1963	Scrapped Crewe Works, December 1963.
46152 *The King's Dragoon Guardsman*	Crewe North	June 1961	Llandudno Junction	September 1961	
46154 *The Hussar*	Edge Hill	July 1961	Llandudno Junction	September 1961	
46156 *The South Wales Borderer*	Holyhead	June 1963	Annesley	November 1963	
46158 *The Loyal Regiment*	Bushbury	December 1960	Trafford Park	December 1960	
46159 *The Royal Air Force*	Crewe North	January 1961	Withdrawal	December 1962	Retained the first style of BR emblem on the tender at withdrawal. Scrapped at Crewe Works, March 1963.
46163 *Civil Service Rifleman*	Holyhead	June 1962	Annesley	January 1963	
46167 *The Hertfordshire Regiment*	Holyhead	June 1963	Annesley	September 1963	
46169 *The Boy Scout*	Crewe North	April 1962	Annesley	January 1963	

Appendix 9: Longsight Allocation

Loco	From	Date	To	Date	Comments
45522 *Prestatyn*	Newton Heath	June 1963	Withdrawn	September 1964	9A was short of power for summer extras. Then worked out of Trafford Park (unofficially) and worked the 5.22 commuter train to Buxton and return the following morning. After withdrawal stored at Buxton shed, September 1964 to 29 May 1965. Scrapped at Central

					Wagon Co., Ince, Wigan, June 1965.
45736 *Phoenix*	Holyhead	June 1960	Crewe North	August 1960	
46106 *Gordon Highlander*	Upperby	June 1958	Newton Heath	April 1960	
46108 *Seaforth Highlander*	Holbeck	June 1952	Preston	November 1959	
46110 *Grenadier Guardsman*	Edge Hill	December 1961	Edge Hill	December 1961	Only on loan for two weeks.
46111 *Royal Fusilier*	Crewe North	June 1953	Holyhead	August 1953	
46111 *Royal Fusilier*	Holyhead	September 1953	Holyhead	May 1957	
46111 *Royal Fusilier*	Holyhead	June 1957	Crewe North	September 1960	
46114 *Coldstream Guardsman*	Edge Hill	February 1956	Edge Hill	February 1956	Only at Edge Hill for two weeks.
46115 *Scots Guardsman*	Upperby	October 1949	Crewe North	September 1960	
46115 *Scots Guardsman*	Upperby	July 1961	Springs Branch	June 1964	
46119 *Lancashire Fusilier*	Crewe North	November 1948	Crewe North	November 1948	On loan for two weeks.
46120 *Royal Inniskilling Fusilier*	Crewe Works	November 1944	Holyhead	June 1954	Loaned to Derby on three occasions for testing purposes.
46120 *Royal Inniskilling Fusilier*	Holyhead	July 1954	Crewe North	June 1956	Loaned to Derby for further testing.
46120 *Royal Inniskilling Fusilier*	Crewe North	June 1960	Camden	September 1960	
46121 *Highland Light Infantry, City of Glasgow Regiment*	Camden	October 1946	Polmadie	June 1949	
46122 *Royal Ulster Rifleman*	Camden	October 1946	Trafford Park	April 1959	
46122 *Royal Ulster Rifleman*	Trafford Park	June 1959	Bushbury	November 1959	
46127 *Old Contemptibles*	Crewe North	June 1960	Crewe North	September 1960	
46128 *The Lovat Scouts*	Crewe North	November 1954	Crewe North	March 1955	
46129 *The Scottish Horse*	Edge Hill	April 1947	Holyhead	February 1953	
46129 *The Scottish Horse*	Crewe North	September 1962	Withdrawn	June 1964	Stored at Longsight, June 1964 to October 1964. Scrapped at Central Wagon Co., Ince, Wigan. November 1964.
46130 *The West Yorkshire Regiment*	Crewe North	June 1953	Upperby	June 1955	
46131 *The Royal Warwickshire Regiment*	Crewe North	April 1947	Camden	October 1947	
46131 *The Royal Warwickshire Regiment*	Camden	November 1947	Kentish Town	October 1957	
46131 *The Royal Warwickshire Regiment*	Crewe North	October 1958	Llandudno Junction	March 1962	
46135 *The East Lancashire Regiment*	Crewe North	September 1956	Crewe North	February 1957	
46135 *The East Lancashire Regiment*	Crewe North	June 1960	Crewe North	September 1960	
46136 *The Border Regiment*	Upperby	March 1957	Upperby	March 1957	
46137 *The Prince of*	Crewe North	November	Crewe North	February	

Wales's Volunteers (South Lancashire)		1956		1957	
46137 *The Prince of Wales's Volunteers (South Lancashire)*	Upperby	October 1957	Edge Hill	October 1957	Two weeks at Longsight.
46137 *The Prince of Wales's Volunteers (South Lancashire)*	Edge Hill	December 1957	Newton Heath	April 1960	
46138 *The London Irish Rifleman*	Crewe North	November 1954	Crewe North	January 1955	
46140 *The King's Royal Rifle Corps*	Crewe North	June 1954	Kentish Town	September 1959	
46140 *The King's Royal Rifle Corps*	Newton Heath	June 1963	Carlisle	September 1964	
46141 *The North Staffordshire Regiment*	Camden	February 1951	Camden	March 1951	
46142 *The York and Lancaster Regiment*	Newton Heath	June 1963	Withdrawn	January 1964`	Stored at Crewe Works, December 1963 and scrapped at Crewe, January 1964. Was the loco in for a repair and then condemned?
46143 *The South Staffordshire Regiment*	Crewe Works	June 1949	Crewe North	February 1951	
46143 *The South Staffordshire Regiment*	Crewe North	December 1952	Holbeck	October 1955	
46143 *The South Staffordshire Regiment*	Holbeck	November 1955	Bushbury	November 1959	
46145 *The Duke of Wellington's Regiment (West Riding)*	Crewe North	April 1947	Camden	November 1949	
46145 *The Duke of Wellington's Regiment (West Riding)*	Camden	December 1949	Holyhead	July 1951	On loan.
46145 *The Duke of Wellington's Regiment (West Riding)*	Holyhead	September 1951	Holyhead	October 1951	
46145 *The Duke of Wellington's Regiment (West Riding)*	Holyhead	January 1951	Holbeck	January 1953	
46147 *The Northamptonshire Regiment*	Edge Hill	September 1956	Holyhead	September 1956	Only two weeks at Longsight.
46148 *The Manchester Regiment*	Upperby	January 1959	Kentish Town	February 1959	
46149 *The Middlesex Regiment*	Camden	October 1946	Holyhead	December 1948	On loan to Holyhead.
46149 *The Middlesex Regiment*	Holyhead	January 1949	Edge Hill	January 1953	
46149 *The Middlesex Regiment*	Edge Hill	June 1956	Holyhead	September 1956	
46149 *The Middlesex Regiment*	Crewe North	June 1961	Withdrawn	September 1963	Withdrawn whilst undergoing repair at Longsight.
46150 *The Life Guardsman*	Bushbury	April 1947	Holyhead	July 1951	
46151 *The Royal Horse Guardsman*	Crewe North	May 1958	Crewe North	June 1958	
46151 *The Royal Horse Guardsman*	Crewe North	July 1959	Millhouses	February 1960	
46152 *The King's Dragoon Guardsman*	Camden	February 1951	Edge Hill	January 1953	
46153 *The Royal Dragoon*	Edge Hill	July 1955	Bushbury	November 1959	
46156 *The South Wales Borderer*	Llandudno Junction	February 1960	Llandudno	March 1960	

46157 *The Royal Artilleryman*	Crewe North	February 1949	Edge Hill	March 1949	Only on loan to Longsight.
46157 *The Royal Artilleryman*	Camden	September 1954	Edge Hill	October 1954	
46158 *The Loyal Regiment*	Edge Hill	July 1955	Kentish Town	April 1959	
46158 *The Loyal Regiment*	Kentish Town	March 1959	Bushbury	November 1959	
46160 *Queen Victoria's Rifleman*	Holyhead	May 1947	Kentish Town	September 1959	
46161 *King's Own*	Holyhead	June 1953	Holyhead	July 1953	
46161 *King's Own*	Holyhead	October 1953	Camden	June 1956	
46162 *Queen's Westminster Rifleman*	Camden	January 1949	Camden	April 1949	
46166 *London Rifle Brigade*	Crewe North	January 1950	Crewe North	February 1950	
46166 *London Rifle Brigade*	Crewe North	November 1954	Crewe North	January 1955	
46166 *London Rifle Brigade*	Crewe North	June 1958	Crewe North	September 1960	
46167 *The Hertfordshire Regiment*	Crewe North	October 1949	Crewe North	December 1951	
46169 *The Boy Scout*	Camden	October 1946	Crewe North	July 1959	
6170 *British Legion*	Crewe North	November 1935	Crewe North	February 1937	
6170 *British Legion*	Crewe North	November 1938	Crewe North	July 1939	

Appendix 10: Bushbury Allocation

Loco	From	Date	To	Date	Comments
45512 *Bunsen*	Crewe Works	October 1948	Upperby	May 1949	Loco rebuilt in July, but not allocated until October.
5514 *Holyhead*	Crewe North	July 1947	Holyhead	August 1948	
45514 *Holyhead*	Holyhead	October 1948	Upperby	May 1949	
45522 *Prestatyn*	Longsight	October 1949	Camden	June 1950	
45525 *Colwyn Bay*	Crewe North	October 1948	Upperby	May 1949	
45525 *Colwyn Bay*	Upperby	October 1949	Upperby	June 1950	
45526 *Morecambe and Heysham*	Edge Hill	October 1949	Upperby	June 1950	
45527 *Southport*	Edge Hill	March 1961	Holyhead	June 1961	
45528 *REME*	Crewe Works	August 1947	Crewe North	June 1948	Not named until 1959.
45529 *Stephenson*	Crewe Works	July 1947	Camden	June 1948	
45529 *Stephenson*	Camden	October 1948	Crewe North	May 1949	
45531 *Sir Frederick Harrison*	Crewe Works	December 1947	Camden	May 1949	
45531 *Sir Frederick Harrison*	Camden	October 1949	Edge Hill	June 1950	
45532 *Illustrious*	Holyhead	August 1948	Camden	May 1949	
45534 *E. Tootal Broadhurst*	Crewe North	October 1949	Longsight	June 1950	

45536 *Private W. Wood, V.C.*	Longsight	October 1949	Longsight	June 1950	
45540 *Sir Robert Turnbull*	Crewe Works	November 1947	Willesden	May 1949	
45540 *Sir Robert Turnbull*	Willesden	October 1949	Longsight	June 1950	
45540 *Sir Robert Turnbull*	Longsight	November 1959	Trafford Park	January 1961	
46114 *Coldstream Guardsman*	Willesden	March 1961	Llandudno	June 1961	
46122 *Royal Ulster Rifleman*	Longsight	November 1959	Willesden	December 1960	
46141 *The North Staffordshire Regiment*	Upperby	November 1959	Willesden	December 1960	
46143 *The South Staffordshire Regiment*	Longsight	November 1959	Trafford Park	December 1960	
46153 *The Royal Dragoon*	Longsight	November 1959	Trafford Park	January 1960	
46158 *The Loyal Regiment*	Longsight	November 1959	Willesden	December 1960	
46161 *King's Own*	Preston	April 1961	Preston	June 1961	
46167 *The Hertfordshire Regiment*	Upperby	April 1961	Preston	June 1961	

Appendix 11: Carlisle Kingmoor Allocation

Loco	From	Date	To	Date	Comments
45512 *Bunsen*	Upperby	November 1964	Withdrawal	March 1965	
45527 *Southport*	Willesden	September 1963	Upperby	August 1964	
45527 *Southport*	Upperby	September 1964	Withdrawn	December 1964	Stored Kingmoor, December 1964 to February 1965 and scrapped at West of Scotland Shipbreaking, Troon, April 1965
45530 *Sir Frank Ree*	Holyhead	December 1964	Withdrawn	December 1965	Stored Kingmoor, December 1965 to March 1966 and scrapped at Motherwell Machinery between March 1966 and July 1966.
45531 *Sir Frederick Harrison*	Upperby	August 1964	Withdrawn	November 1965	Stored Kingmoor, November 1965 to January 1966 and scrapped at Campbells, Airdrie, between January and March 1966.
45535 *Sir Herbert Walker, K.C.B.*	Edge Hill	October 1962	Withdrawn	November 1963	Stored November 1963 to February 1964 at Kingmoor, then at 16B Annesley, February 1964 to August 1964 on its way to be scrapped at Rigley and Sons, Bulwell, September 1964.
45736 *Phoenix*	Upperby	September 1964	Withdrawal	September 1964	Stored Kingmoor, September 1964 to December 1964 and scrapped at Hughes Bolckows, North Blyth, January 1965.
46110 *Grenadier Guardsman*	Edge Hill	June 1963	Withdrawn	February 1964	Stored Upperby, February 1964 to November 1964 and scrapped at McWilliams, Shettlestone, December 1964.
46115 *Scots Guardsman*	Springs Branch	July 1964	Withdrawn	January 1966	Stored Kingmoor, January 1966 to August 1966 and then moved to Keighley and Worth Valley for preservation.
46116 *Irish Guardsman*	Edge Hill	October 1962	Withdrawn	September 1963	Stored Crewe Works, September 1963 and scrapped there in September 1963.
46124 *London Scottish*	Edge Hill	October 1962	Withdrawn	December 1962	Stored Kingmoor, December 1962 to April 1963 and scrapped at Crewe Works, April 1963.

46128 *The Lovat Scouts*	Springs Branch	September 1962	Withdrawn	May 1965	Stored Kingmoor, May 1965 to July 1965and scrapped at Motherwell Machinery, Wishaw, July 1965.
46129 *The Scottish Horse*	Crewe North	July 1961	Upperby	September 1961	
46132 *The King's Regiment Liverpool*	Upperby	September 1963	Withdrawn	February 1964	Stored February 1964 to February 1965 at Kingmoor and scrapped at West of England Shipbreaking, Troon, in April 1965.
46140 *The King's Royal Rifle Corps*	Longsight	June 1963	Withdrawn	November 1965	Stored at Kingmoor, November 1965 to February 1966 and then scrapped at McWilliams,Shettlestone, March 1966.
46152 *The King's Dragoon Guardsman*	Holyhead	January 1965	Withdrawn	April 1965	Stored Kingmoor, April 1965 to June 1965 and scrapped Motherwell Machinery, Wishaw, July 1965.
46155 *The Lancer*	Crewe North	September 1964	Withdrawal	November 1964	Stored Kingmoor, November 1964 to January 1965 and scrapped at West of Scotland Shipbreaking, Troon, February 1965.
46157 *The Royal Artilleryman*	Upperby	June 1963	Withdrawn	January 1964	Stored Crewe Works, January 1964 and scrapped there in February 1964.
46160 *Queen Victoria's Rifleman*	Upperby	June 1963	Withdrawal	May 1965	Stored Kingmoor, May 1965 to June 1965 and scrapped at Motherwell Machinery, Wishaw, July 1965.
46162 *Queen's Westminster Rifleman*	Upperby	July 1962	Withdrawal	June 1964	Stored Kingmoor, throughout August 1964, and then scrapped at Connels, Calder, September 1964.
46166 *London Rifle Brigade*	Upperby	September 1963	Withdrawal	October 1964	Stored at Kingmoor, October to November 1964 and scrapped at West of England Shipbreaking, Troon, December 1964. Engine was observed with a 12B Upperby shed plate in September 1964.

Appendix 12: Carlisle Upperby Allocation

Loco	From	Date	To	Date	Comments
45532 *Illustrious*	Saltley	June 1962	Withdrawn	February 1964	Stored February 1964 to December 1964 at Upperby. Scrapped Campbells, Airdrie, January1965.
45540 *Sir Robert Turnbull*	Saltley	June 1962	Withdrawn	May 1963	Stored at Crewe Works, June 1963 and scrapped at Crewe Works, July 1963.
45736 *Phoenix*	Willesden	July 1964	Kingmoor	September 1964	
46103 *Royal Scots Fusilier*	Saltley	June 1962	Holbeck	September 1962	
46106 *Gordon Highlander*	Saltley	June 1962	Withdrawn	December 1962	Stored at Upperby, October 1963 to February 1963, but reinstated to traffic for a few weeks in late January 1963, before being stored at Kingmoor in March 1963 and scrapped at Crewe Works, April 1963.
46108 *Seaforth Highlander*	Longsight	November 1960	Withdrawn	January 1963	Stored Kingmoor, February 1963 to April 1963 and scrapped at Crewe Works, May 1963.
46115 *Scots Guardsman*	Crewe North	March 1949	Longsight	October 1949	
46115 *Scots Guardsman*	Crewe North	May 1961	Longsight	July 1961	
46116 *Irish Guardsman*	Holyhead	June 1956	Kentish Town	October 1957	

CARLISLE UPPERBY ALLOCATION ■

46118 *Royal Welch Fusilier*	Saltley	June 1962	Withdrawn	June 1964	Stored at Upperby, June 1964 to October 1964 and scrapped at Connels, Calder, November 1964.
46122 *Royal Ulster Rifleman*	Saltley	June 1962	Annesley	December 1962	
46122 *Royal Ulster Rifleman*	Annesley	October 1964	Withdrawn	October 1964	There is some doubt if the engine reached Upperby, but it was reported as stored at Upperby between October 1964 and January 1965 and scrapped at Draper's, Hull, in February 1965.
46123 *Royal Irish Fusilier*	Saltley	June 1962	Withdrawn	October 1962	Stored at Upperby, October 1962 to March 1963, before being scrapped at Crewe Works in April 1963.
46125 *3rd Carabinier*	Crewe North	May 1962	Crewe North	May 1962	Only stayed four weeks.
46126 *Royal Army Service Corps*	Holyhead	February 1958	Preston	November 1959	
46127 *Old Contemptibles*	Crewe North	May 1962	Withdrawn	December 1962	After withdrawn, put back into service and seen on North Wales coast in February 1963. Stored October 1962 to April 1963 and scrapped at Crewe, May 1963.
46129 *The Scottish Horse*	Camden	October 1957	Crewe North	November 1957	
46129 *The Scottish Horse*	Kingmoor	September 1961	Crewe North	March 1962	
46130 *The West Yorkshire Regiment*	Longsight	June 1955	Edge Hill	June 1957	
46132 *The King's Regiment Liverpool*	Saltley	June 1962	Kingmoor	September 1963	
46134 *The Cheshire Regiment*	Edge Hill	May 1962	Withdrawn	November 1962	Stored Upperby, December 1962 to April 1963 and scrapped Crewe Works, April 1963.
46135 *The East Lancashire Regiment*	Holyhead	July 1954	Crewe North	September 1956	
46136 *The Border Regiment*	Crewe Works	March 1950	Longsight	March 1957	Only at Longsight for three weeks, probably on loan.
46136 *The Border Regiment*	Longsight	March 1957	Preston	November 1959	
46136 *The Border Regiment*	Preston	September 1960	Crewe North	October 1960	
46136 *The Border Regiment*	Holbeck	September 1962	Withdrawn	April 1964	Scrapped Crewe Works April 1964.
46137 *The Prince of Wales's Volunteers (South Lancashire)*	Saltley	June 1962	Withdrawn	October 1962	Stored at Upperby, October 1962 to February 1963 and then at Kingmoor, March 1963 to April 1963, before being scrapped at Crewe Works, May 1963.
46138 *The London Irish Rifleman*	Holyhead	November 1962	Withdrawn	February 1963	Stored at Crewe Works, February 1963 to April 1963 and scrapped Crewe Works, May 1963.
46141 *The North Staffordshire Regiment*	Holbeck	January 1956	Bushbury	November 1959	
46141 *The North Staffordshire Regiment*	Saltley	July 1962	Withdrawn	April 1964	Stored at Crewe Works, April 1964 to June 1964 and scrapped there in July 1964.
46146 *The Rifle Brigade*	Crewe North	July 1951	Camden	April 1952	
46146 *The Rifle Brigade*	Camden	January 1954	Camden	March 1954	
46147 *The*	Crewe	August 1947	Crewe	August	On loan for three weeks.

Northamptonshire Regiment	North		North	1947	
46147 *The Northamptonshire Regiment*	Edge Hill	March 1949	Camden	April 1952	
46147 *The Northamptonshire Regiment*	Holyhead	October 1957	Holyhead	November 1957	On loan for four weeks.
46148 *The Manchester Regiment*	Camden	October 1957	Longsight	January 1959	Initially on loan.
46148 *The Manchester Regiment*	Kentish Town	April 1959	Millhouses	February 1960	
46155 *The Lancer*	Crewe North	February 1957	Crewe North	February 1957	Probably on loan as only two weeks.
46157 *The Royal Artilleryman*	Saltley	July 1962	Kingmoor	June 1963	
46160 *Queen Victoria's Rifleman*	Saltley	July 1962	Kingmoor	June 1963	
46162 *Queen's Westminster Rifleman*	Saltley	July 1962	Kingmoor	September 1963	
46165 *The Ranger 12th London Regiment*	Crewe Works	July 1952	Rugby Test Centre	December 1955	Being tested at Rugby.
46165 *The Ranger 12th London Regiment*	Rugby Test Centre	January 1956	Rugby Test Centre	January 1956	Only back at Carlisle for two weeks.
46165 *The Ranger 12th London Regiment*	Rugby Test Centre	June 1956	Crewe	August 1956	EHC says Crewe South (4 August 1956).
46165 *The Ranger 12th London Regiment*	Crewe	November 1956	Crewe	February 1957	EHC says Crewe South (1 June February 1957).
46166 *London Rifle Brigade*	Crewe North	September 1962	Kingmoor	September 1963	
46167 *The Hertfordshire Regiment*	Crewe North	October 1956	Preston	April 1959	
46167 *The Hertfordshire Regiment*	Preston	September 1960	Bushbury	March 1961	

RECOMMENDED READING

Cook, Powell, Tuplin and Johnson, *Royal Scots of the LMS* (Ian Allan, 1970).

Goodman, John, *LMS Locomotive Names* (RCTS, 1994). Details the origins and naming policy of the LMS for its locomotive fleet.

Hands, P.B. *'What Happened to Steam'* series (Defiant Publications). This lists the allocation and the disposal details of the rebuilt Patriots and rebuilt Scots. There has been much published questioning of some of the data in these books, and the author has identified a small number of errors on allocations by cross-referencing them with the 'Weekly Movement Sheets' issued by BR, which detail the allocation and loans on a weekly basis. The scrapping data has also been crossed-checked with other sources, but the majority of the rebuilt Class 7s were scrapped at Crewe Works and the author has corroborating evidence for most of the engines not scrapped at Crewe that he has been happy to use as a source.

James, Hunt and Essery, *LMS Locomotive Profiles: No 1. The Rebuilt 'Royal Scots'* (Wild Swan Publications, Langridge, E.A., Under 10 CMEs. Vol. 1 1912–1944 (Oakwood Press, 2011). More a detailed analysis of the mechanical details, with extensive detail drawings and details of livery changes.

Powell, A.J., Stanier *Locomotive Classes* (Ian Allan, 1991).
John Powell, who was a locomotive inspector for the CME Department, wrote extensively in the railway press (such as *Trains Illustrated*) under the nom de plume of '45671' and his reminiscences give a fascinating insight into the 'real world' of steam locomotive workings.

Sixsmith, Ian, *The Book of the Royal Scots* (Irwell Press, 1999).

Sixsmith, Ian, *The Book of the Patriots* (Irwell Press, 2003).

'Steam for Scrap' (Atlantic Transport Publications, 1993). This is a revised version of a series published over a number of years and updates some of the data. The disposal of many locomotives was not documented well and many of the private yards did not keep detailed records. However, the rebuilt Class 7s were mainly scrapped at Crewe Works and the small number scrapped at private yards were reasonably well documented.

The Book of the Patriots, Photographic Accompaniment No: 1 (Irwell Press, 2007).

Townsin, Ray, *The Jubilee 4-6-0s* (RCTS, 2006). An in-depth analysis of the Jubilees, including a good section on the two rebuilt Jubilees.

INDEX

AWS 43
allocations 82–129
A3 Pacifics 89

brake shoes 37

cabs 37
Crewe Works 84, 86, 111, 148, 152, 153, 157
Chapelon 15
Coleman, T F. 15,16
coupling rods 35
Cox, E. S. 12

depots
    Annesley 141–5.
    Bristol Barrow Road 113, 119, 131, 140
    Bushbury 111, 164
    Buxton 163
    Camden 84, 88, 95, 99, 100, 111
    Canklow 123.
    Carlisle (Kingmoor) 87, 113
    Carlisle Upperby 87, 115
    Corkerhill 106,107,156.
    Crewe North 10, 82, 84, 91, 92, 94, 111
    Darnall 124
    Derby 126, 127
    Edge Hill 87, 98, 99
    Holbeck (Leeds) 82, 87–90, 106, 138
    Holyhead 87, 95
    Kentish Town 120, 122, 123
    Leicester (GCR) 134
    Llandudno Junction 95
    Longsight (Manchester) 87, 108, 146
    Low Moor (Bradford) 90, 137, 138

Millhouses (Sheffield) 123
Mirfield 138, 139
Newton Heath 119, 120, 131
Nottingham 25, 126
Polmadie (Glasgow) 87, 104–6
Preston 117, 118
Saltley 139–41
Springs Branch (Wigan) 99, 117, 134–5
Trafford Park (Manchester) 111, 127, 128
Willesden 100

disposal 154, 156–7

Fairburn, C. E. 18
footsteps 40
front buffer beams. 37.
front smoke box number plates 68

Horwich Works 15

lamp brackets 42
Langridge, E. A. 15
Liveries 50–51
    British Railways 53
    LMS 1942–46 51.
    LMS post-1946 52
locomotive exchanges, 1848 78, 80

nameplates 58–63

overhead warning indicators 71

plaques 64–7

RCTS 10
repair categories
   Heavy General 147
   Heavy Intermediate 148
   Light Classified 149
   Light Intermediate 148
   Non-Classified 149
repair locations 146–7
rocking grates 38
roller bearing connecting rods 78
Riddles, R. A. 27
Rowledge, P. 27

sandboxes 41
self-cleaning smoke box 76, 77

smoke deflectors 37, 76
speedometer 43
Stanier, Sir William 12, 13

tender emblem 44
tenders 44
testing 74, 75, 80, 81
top feed 42

wheels 30–35
withdrawal 152–9
works maintenance 147–51

yellow warning stripe 69